The Spirit of Numbers

The Spirit of Numbers

Sacred Mathematics and the Language of Creation

Elodie Varkash

Mindful Pages

Published in 2026

ISBN: 9789375363378 (PB)
ISBN: 9789375365334 (eBook)

Published by

Mindful Pages
Imprint of Alpha Editions LLC
312 W. 2nd St #1834
Casper, WY 82601, USA

This book was developed using advanced writing tools to support the drafting and ideation process. All content has been thoughtfully curated, refined, and reviewed to ensure quality and relevance.

Content

Introduction

A child sits on the floor, legs folded, moving pebbles into a line. One, two, three. The act is unremarkable, almost invisible to adult eyes, yet something decisive is taking place. The world is being sorted. Not owned or mastered, but touched lightly, divided and rejoined through attention. Each stone stands alone, and yet belongs to a sequence that did not exist a moment before. In that small gesture lies one of humanity's oldest powers: the ability to notice quantity and allow it to shape meaning.

Most of us lose sight of that moment. Numbers grow up with us, hardening into grades, deadlines, interest rates, passwords. They arrive wearing the uniform of authority and demand obedience rather than curiosity. We learn to brace ourselves against them, to fear being exposed as incompetent, careless, or insufficiently quick. At the same time, we hand them extraordinary trust. We step onto aircraft calculated to the decimal, swallow medicines dosed by milligram, wake to alarms synchronised with atomic clocks. We rely on numbers with our lives while privately resenting them.

This uneasy relationship is not accidental. It is the result of a long narrowing, a cultural decision to treat number primarily as an instrument of control. What once helped us orient ourselves in the cosmos has been reduced, in many contexts, to a tool for extraction, comparison, and judgement. Yet numbers themselves have not changed. They have been with us since the first cut marks on bone, since fingers traced cycles of moon and tide. Long before they were pressed into spreadsheets, they were sung, walked, prayed, and carved into stone.

Across civilisations, number has been approached as something more than a convenience. Temples rose according to proportion, not only to stand but to resonate. Music was tuned to ratios believed to mirror the order of the heavens.

Myths spoke of creation unfolding through countable stages, of worlds born from measure and balance. Even the stars were described through number, not as cold distances but as participants in a patterned whole. To attend to number was, for much of human history, a way of attending to meaning itself.

This book begins with a simple but unsettling question: what if numbers are not only human inventions for managing reality, but hints of a larger intelligence woven into it? Not intelligence in the narrow sense of calculation, but an ordering principle that invites relationship. What if quantity, far from flattening experience, can deepen it when approached with care? The aim here is neither to mystify mathematics nor to romanticise the past, but to recover a fuller way of seeing that has been obscured by fear and overuse.

The urgency of this recovery is easy to miss because numbers feel omnipresent and neutral. They saturate modern life so completely that we rarely pause to ask what kind of world they are quietly shaping. We measure productivity by output, worth by income, attention by clicks. We count steps, followers, calories, years. These counts influence how we treat ourselves and one another, often without reflection. When numbers become the sole language of value, they begin to dictate rather than describe. The result is a culture both obsessed with measurement and starved of meaning.

At the same time, the scientific story of the universe has never been more astonishing. Simple equations capture the fall of apples and the dance of galaxies. Fixed constants hold reality together with uncanny precision. Patterns repeat from the spiral of a shell to the curve of a hurricane, from branching trees to the structures of lungs and rivers. We live in a cosmos that is mathematically legible, at least in part, and that legibility raises profound questions. Are we discovering a script already written, or imposing our own grammar on an indifferent world? Why does mathematics work at all?

These questions are not confined to laboratories or lecture halls. They touch the most intimate corners of human life. Anyone who has counted breaths to steady anxiety, marked days in grief, or repeated a prayer bead by bead knows that number can soothe as well as constrain. Anyone who has been reduced to a score, a rank, or a statistic knows that number can wound. The same symbols that guide spacecraft can also become instruments of shame or superstition. Our task is not to choose between reverence and rigour, but to learn how to hold both without distortion.

The organising metaphor of this book is language. Numbers will be treated not as mute objects but as a grammar through which reality speaks and through which we respond. Like any language, mathematics has vocabulary, syntax, and style. It can be used to command or to converse, to dominate or to listen. We can shout in it, or we can learn its subtleties. Approached this way, numeracy becomes less about competence and more about relationship, less about answers and more about attention.

The chapters that follow move deliberately from the intimate to the expansive. We begin where everyone begins, with counting as a bodily act. The child lining up toys, the adult counting steps or breaths, the simple reassurance that comes from order emerging out of scatter. From there, we meet figures such as Pythagoras, not as remote geniuses but as listeners attuned to harmony in waves and hammers. Their insights arose from wonder before they hardened into doctrine.

We trace the long journey from tally marks scratched into bone to full numeral systems capable of tracking trade, time, and obligation. This practical history matters because it grounds later symbolism in lived necessity. Numbers first helped us remember, coordinate, and trust. Only later did they acquire the aura of abstraction that now intimidates so many.

Understanding this shift helps explain why modern adults can feel both dependent on numbers and alienated from them.

Harmony becomes a central theme as ratios move from music into bodies and buildings. A string divided just so produces consonance; a room shaped with care invites ease. Proportion is felt before it is calculated. When measurement takes over from harmony, something is gained in precision and something is lost in resonance. The book does not mourn this change, but examines it closely, asking how utility and beauty might be reunited.

As the grammar of quantity unfolds, different kinds of numbers appear as characters rather than hurdles. Zero, negative, fractional, infinite. Each carries its own intuition and its own disturbance. Infinity, in particular, stands at the edge of thought, provoking awe rather than mastery. The ancient question of whether numbers are discovered or invented re-emerges here, not as an academic puzzle but as a live issue that shapes how we read the world.

Symbolic numbers across cultures reveal another layer of meaning. One and two, three and four, seven and twelve. These patterns recur not because of secret codes, but because they resonate with human experience of unity, division, movement, and time. The line between symbolism and superstition is examined carefully, with an insistence that true symbols open reflection rather than closing it down.

Geometry then takes centre stage, not as a branch of mathematics to be mastered, but as a visual theology. Circles, triangles, crossings, stars. Shapes that gather bodies, orient worship, and tell stories without words. Sacred architecture teaches proportion through light and movement, long before it is explained. Even the simple act of drawing a pattern can become a form of devotion when approached with patience.

The living world offers its own lessons. Spirals in shells, branching in trees, waves in water and crowds. Pattern emerges from apparent chaos and dissolves back into it. Human beings are storytellers about pattern, sometimes too eager, sometimes too dismissive. Learning to notice without forcing meaning becomes a quiet discipline, a way of praying with the eyes open.

Not all encounters with number are gentle. The book does not shy away from the ways numbers hurt. Metrics can tyrannise, superstition can trap, probability can deceive. Modern life provides countless examples of counting gone wrong, of quantity replacing judgement and compassion. Healing our relationship with numbers requires boundaries as well as insight, a willingness to let some things remain unmeasured.

Science returns the discussion to a larger scale. Elegant laws, stable constants, and the persistent mystery of why mathematics describes nature so well. Here humility is essential. The history of physics is full of corrected errors and provisional truths. To read the script of nature is to accept that some pages remain illegible, and that this does not diminish their reality.

The final movements of the book return number to daily life as contemplation. Counted breath, repeated words, small tallies of gratitude. These practices are modest by design. They resist the grandiosity that so often infects spiritual discourse. Letting go of the last digit, accepting approximation, becomes an ethical stance as much as a practical one.

Living by a different measure is the quiet challenge that remains. Success, time, community. Each can be counted in ways that distort or in ways that illuminate. The invitation is not to abandon numbers, but to choose our scales consciously, aware of what they encourage and what they obscure. To walk through the world as a reader of patterns, attentive but not enslaved.

This book does not promise secret knowledge or universal answers. It offers something rarer and more demanding: a way of paying attention. It asks whether we can learn to meet numbers with the same curiosity we bring to music, art, or prayer. Whether quantity can become a site of encounter rather than anxiety. Whether the language we use to describe the world might also teach us how to belong within it.

The pages ahead explore these questions slowly, deliberately, and without shortcuts. They assume intelligence, patience, and a willingness to sit with uncertainty. If numbers have come to feel like enemies or abstractions, this book invites you to meet them again, as companions pointing beyond themselves. Not to mastery, but to participation in an order that is at once precise and alive.

Chapter 1
When Counting Becomes Wonder

A child sits on the carpet, absorbed, arranging toys into a careful line. There is no audience and no instruction. The movement is slow, deliberate. Each object is placed, adjusted, sometimes removed and returned. One, two, three. The counting is not yet precise, but that hardly matters. What matters is the seriousness of attention. Something in the child recognises that the world can be gathered, that difference can be held without confusion, that order is not imposed but discovered through touch.

Adults rarely remember this moment, though almost everyone lived it. Before numbers became problems to solve or hurdles to clear, they arrived as companions to play. Stones were counted not to pass a test but to reassure. Steps were marked not to optimise performance but to make distance bearable. Breaths were numbered not to control the body but to stay with it. Counting began as a way of staying oriented in a world that was larger, louder, and more unpredictable than a small person could otherwise manage.

The paradox is stark. The same activity that once calmed and focused us now provokes unease. Many grown people confess, without irony, to being "bad with numbers", even as they organise their lives around them. We trust quantities to tell us whether we are healthy, solvent, productive, or late. We glance at statistics for comfort and recoil when they threaten certainty. Counting has become both a refuge and a source of anxiety, a language we depend on but rarely enjoy speaking.

This chapter returns to the beginning, not out of nostalgia, but because something essential was present there that has been lost along the way. Counting did not begin as abstraction. It began as ritual. The repeated placement of objects, the marking of time, the steady naming of more and less wer ways of making sense of experience. To count was to say: this moment follows that one; this thing is here, and so is the next. Order emerged not through authority, but through attention.

That quality of attention matters. The child lining up toys is not trying to dominate the world, only to relate to it. Each number spoken is an act of recognition. This is one. This is another. They are not the same, yet they belong together. Long before mathematics became a discipline, counting taught a basic lesson about reality itself: difference does not destroy unity; it makes it intelligible. That lesson will echo throughout this book.

From this domestic scene, the chapter widens its gaze to a shoreline thousands of years away. Pythagoras, remembered now mostly as a theorem, was first and foremost a listener. The stories that survive place him not at a desk but among sounds: hammers striking anvils, strings vibrating, waves breaking in rhythmic intervals. He noticed that harmony followed number, that simple ratios produced consonance while others jarred. This was not the discovery of a formula in the modern sense, but the recognition of a pattern that invited contemplation.

It is easy to forget how radical this was. To suggest that the world could be understood through number was not to strip it of mystery, but to deepen it. Harmony was evidence of an underlying order that could be sensed before it was calculated. Pythagoras did not begin with examinations or proofs. He began with wonder. Mathematics, in this telling, emerges not as a cold instrument but as a way of listening more carefully to what is already there.

This shift from story to reflection matters because it unsettles a modern assumption: that mathematics is primarily about correctness. In its earliest expressions, number was about resonance. It described relationships that could be heard and felt. A lyre string shortened by half did not merely demonstrate a ratio; it produced an octave that stirred the body. The pleasure was inseparable from the insight. To learn number was to participate in a harmony that exceeded the self.

Such participation has practical roots. The chapter traces how counting moved from gesture and sound into mark and system. Scratches on bone, tallies on wood, knots on cord. These early records were not philosophical abstractions but tools for survival. They tracked animals, debts, seasons. They allowed small groups to remember more than any one person could hold. From these humble beginnings grew calendars that coordinated planting and harvest, and numerical systems that enabled trade across distance.

This evolution grounds later symbolism in necessity. Numbers did not float into human culture as metaphors; they were pressed into service. A tally mattered because hunger mattered. A calendar mattered because winter came whether one was ready or not. By following this path from mark to system, the chapter resists the temptation to treat sacred number as something separate from daily life. Reverence grew out of need, not in spite of it.

As systems developed, so did power. To count is to make claims: how much, how many, how long. Those claims could be contested. Numbers became instruments of agreement and, at times, of coercion. A debt recorded in marks carried weight. A census reshaped communities. With systematisation came distance. Counting was no longer always done by the person who felt its consequences. This distance is one reason numbers later acquired their reputation for coldness.

The modern inheritance of this history is ambivalent. We live inside vast numerical systems we did not design and do not fully understand. We trust them because we must. A bank statement reassures precisely because it appears objective. A medical statistic steadies the nerves even as it frightens. Quantity offers clarity in situations charged with emotion, and that clarity is seductive. Yet the same numbers can also overwhelm, reducing complex lives to probabilities and scores.

This chapter does not moralise that tension; it examines it. Why do so many adults flinch when confronted with numbers, even as they reach for them instinctively in moments of uncertainty? Part of the answer lies in education, where numeracy is often framed as performance rather than practice. Another part lies in culture, which has trained us to see numbers as verdicts rather than invitations. Once quantity is treated as judgement, curiosity withers.

The hunger beneath the fear is for certainty. Numbers promise firmness in a world that feels increasingly unstable. They appear to offer ground that does not shift. When that promise is fulfilled, we relax. When it fails, we panic. The relationship becomes brittle. What began as a ritual of reassurance turns into a demand for control. The chapter traces this psychological arc carefully, showing how it arises and why it matters.

Yet fear is not the only possible response. The same numerical encounters that intimidate can, under different conditions, invite reflection. A bill can prompt questions about value. A statistic can open a conversation about risk and care. The move the chapter proposes is subtle but consequential: from treating numbers as authorities to treating them as interlocutors. That move restores a measure of agency without denying the real power of quantity.

Throughout, a recurring image threads the discussion: the line. Toys lined up on the floor. Waves breaking in sequence. Tally marks etched one after another. Bills itemised in columns. Lines organise experience. They create before and after, more and less. They can calm or constrain. By attending to how lines function in different contexts, the chapter invites readers to notice the quiet structures shaping their own thinking.

Modern relevance presses in from all sides. The world now counts at a scale unimaginable to earlier cultures. Data streams measure behaviour in real time. Algorithms translate human actions into patterns of prediction. The ancient act of counting has been amplified beyond recognition. In this context, returning to the origins of numeracy is not an antiquarian exercise; it is a form of discernment. Without a felt sense of what counting is for, we risk being counted rather than counting.

The chapter's scope is deliberately broad, but its movement is careful. It does not attempt to explain mathematics, nor to catalogue historical developments exhaustively. Instead, it follows a single thread: how counting shapes our sense of order, and how that shaping can tilt towards wonder or towards fear. By moving between intimate scenes and historical moments, it shows that this question has always been with us.

What emerges is a reframing of numeracy itself. Not as a school subject, nor as a professional skill, but as a basic human capacity for attention. To count is to pause long enough to notice difference and connection. That pause is increasingly rare. Reclaiming it is an act of quiet resistance against haste and abstraction. It is also an opening into the larger themes of the book, where number will appear again and again as a bridge between the measurable and the meaningful.

The chapter ends by inviting readers to reconsider their own first encounters with number. Not the classroom or the exam hall, but the floor, the path, the breath. From there, the story moves forward, carrying with it the possibility that counting need not be a burden. It can be, once again, a way of standing in the world with steadiness and curiosity, attentive to pattern without being ruled by it. The wonder that begins in childhood has not vanished; it has only been covered over. The work ahead is to uncover it, patiently, one number at a time.

The child who learns to count

On a cold January morning in 1907, in a modest apartment block on Via dei Marsi in Rome, Maria Montessori watched children move small wooden cylinders in and out of their slots. No one was correcting them. No one was counting aloud on command. Yet patterns emerged. Objects were aligned, rearranged, grouped, returned. Montessori recorded the scene carefully, noting how children repeated these acts with a seriousness usually reserved for adult work. She did not call it mathematics. She called it concentration. Order arrives without instruction here.

The scene matters because it fixes a misunderstanding that still shadows numeracy. Counting does not begin as compliance. It begins as attention. Long before a child encounters numerals on a page, they encounter quantity with their hands and eyes. Stones are gathered. Steps are paced. Breaths are noticed. The child lining up toys is not rehearsing a future exam; they are testing a hypothesis about the world: that difference can be held together without confusion.

This is not sentimentality. It is observation. Montessori's notebooks from Rome in 1907 describe children who persisted with ordering tasks far beyond what any reward system could explain. The behaviour was measurable. Children returned to the same sequence dozens of times. They corrected errors themselves. They appeared calmer afterwards. These were visible outcomes, not theories. Counting, in this register, functioned as a basic human ritual for establishing orientation.

The ritual quality becomes clearer when we slow the scene down. A toy is placed. Another follows. The line grows. Each placement creates a small pause, a moment of judgement. Is this next? Does it belong? The child measures with the body before measuring with symbols. The metric is tactile: reach, spacing, balance. Trust is built through repetition, not explanation. The implication is direct. Numeracy begins as a way of settling into space and time, not as a way of demonstrating correctness.

That claim can be named plainly. Early counting shapes a sense of order. The explanation is not neurological mystique but behaviour. Repeated sequencing reduces uncertainty. It gives a child a reliable procedure when the environment feels large. The implication reaches beyond childhood. Adults who count steps when anxious, or breaths when overwhelmed, are drawing on the same mechanism. The behaviour persists because it works.

History corroborates the point. The Ishango bone, discovered in the 1950s near Lake Edward in what is now the Democratic Republic of the Congo and dated to roughly 20,000 years ago, bears a series of notches arranged in deliberate groupings. Archaeologists disagree on the precise function, but they agree on one thing: the marks are not random. Someone counted. The place is specific. The date is checkable. The behaviour is clear. Long before writing, humans marked quantity to stabilise memory.

Show the act, then name it. Counting externalises time. It allows experience to be revisited. A notch stands in for a moment that would otherwise vanish. The explanation does not require grand symbolism. Hunters needed to remember cycles. Gatherers needed to track yield. The implication is that number emerged from survival, not abstraction. Sacred meaning, where it later appeared, was layered onto a practice already trusted.

The line returns here, scratched into bone rather than laid on a carpet. The motif holds. Lines separate and connect. They impose sequence without erasing difference. When marks are grouped, a pattern is proposed. The maker commits to a structure that can be checked later. This is the quiet origin of accountability. A tally allows a claim to be tested. That is a social achievement as much as a cognitive one.

Move forward a few millennia and the same logic governs calendars. In Mesopotamia, by the late fourth millennium BCE, lunar cycles were marked to coordinate planting and labour. Clay tablets from Uruk record rations distributed by count. The behaviour is visible: measures of barley, days worked, mouths fed. Trust is mediated by quantity. Efficiency increases because obligations can be tracked. The implication is sobering. Once counting enters collective life, it shapes power.

Yet even here, the ritual dimension remains. To count days is to acknowledge limits. A month ends whether one wishes it to or not. Counting offers predictability, but it also imposes patience. This double edge persists. Quantity reassures because it constrains. It draws a line around what can be expected.

Children intuit this. When a child counts steps to the door, the distance becomes bearable. The behaviour is simple. The abstraction is safety. The explanation is procedural. By breaking a span into units, the child gains agency. The implication is that counting teaches endurance as much as accuracy.

The educational system often obscures this origin. By the time numeracy appears on a timetable, it has been stripped of its bodily cues. Numbers arrive as symbols divorced from the behaviours that once gave them sense. Performance replaces practice. Correctness replaces orientation. The result is measurable. Anxiety rises. Avoidance follows. Adults later report being "bad at maths" without being able to name a specific failure. The fear is general because the rupture was early.

This is the limitation that must be acknowledged. Counting can be taught in ways that erode trust. Standardised testing regimes, introduced widely in the United Kingdom from the late twentieth century, tie numeracy to ranking. The metric is score. The behaviour becomes comparison. The implication is predictable. When numbers are experienced primarily as judgements, curiosity contracts. This counter-example matters because it explains the adult flinch.

Reconciliation requires returning to the ritual without discarding the system. Montessori herself insisted on this balance. Her materials were precise. Errors were visible. Yet the child retained control of pace and sequence. The named example shows a path forward. Structure need not eliminate wonder if it preserves agency.

The same balance appears in another documented practice. In the Japanese tradition of *shugyō*, particularly in Zen monasteries from the Kamakura period onward, novices counted breaths as part of seated meditation. The behaviour is recorded in manuals dated to the thirteenth century. The metric is simple. One to ten, repeated. The purpose is not calculation but steadiness. When attention wanders, the count resets. The implication is that number can discipline without coercing.

This is not an argument for importing spiritual techniques into classrooms. It is an observation about continuity. Counting as ritual crosses cultures because it addresses a common need: the need to stay present in duration. The line of numbers holds the mind where it might otherwise scatter.

The motif reappears in domestic life. Watch a child set the table. Plates are counted because places matter. Someone will sit here. Another there. The behaviour assigns belonging. The abstraction is community. The explanation is spatial. The implication extends. When adults count guests, seats, portions, they are doing more than logistics. They are rehearsing care.

Modern professionals recognise a degraded version of this when they track tasks. Checklists calm because they externalise obligation. The metric is completion. The behaviour is ticking off. The abstraction is control. The explanation is cognitive load reduction, described in psychology literature by George Miller's work on memory limits in 1956. The implication is that counting persists because it reduces strain.

Yet the risk remains. When the list becomes the point, meaning thins. Counting that once served orientation becomes an end in itself. This is where fear enters. Bills arrive itemised. Medical reports enumerate risks. The behaviour shifts from arranging to bracing. The same line that once reassured now threatens.

The difference lies in posture. The child leaning over toys counts forward. The anxious adult scans backward, searching for error. The metric is identical. The experience is not. This distinction matters because it suggests that the problem is not number, but relationship.

Return to the opening image. The toys do not line themselves. The child commits to the line, adjusts it, sometimes abandons it. There is play in the seriousness. That play is not frivolous. It is exploratory. It allows failure without penalty. Counting retains its power because it tolerates revision.

Historical evidence supports this tolerance. Early merchant tallies were often split sticks, one half kept by buyer, one by seller. Errors could be reconciled by fitting pieces together. The behaviour acknowledged fallibility. The implication was trust maintained through shared structure. Counting did not eliminate dispute; it provided a way to address it.

The line appears again, broken and rejoined. The motif holds.

What, then, is at stake in reframing numeracy as ritual? The claim is modest but consequential. If counting is recognised as a way of attending, not merely of evaluating, then fear loosens. Adults can revisit the behaviour without the baggage of performance. Professionals can ask not only whether a number is correct, but what behaviour it is shaping.

The explanation does not deny the need for precision. Bridges must hold. Doses must be accurate. The implication is contextual. Precision serves a purpose; it does not define the whole relationship. When counting forgets its origin in orientation, it becomes brittle. When it remembers, it regains flexibility.

This subchapter does not resolve the tension. It sets a test. The next time a number appears—on a screen, a page, a receipt—what line is being drawn, and for whom? Is the count inviting participation or demanding submission? The answer will not be theoretical. It will be behavioural.

Pythagoras at the shore

In the late sixth century BCE, on the coast of southern Italy near Croton, a Greek émigré stood listening. Pythagoras of Samos had arrived there around 530 BCE, after years of travel through the eastern Mediterranean. Croton was prosperous, loud with trade and metalwork, and close to the sea. According to later testimony preserved by Aristotle and Iamblichus, it was here that Pythagoras began teaching that number lay beneath the order of things. The claim endured. The scene that lingers is quieter: waves breaking in intervals, hammers striking anvils, strings vibrating in air.

The historical record is fragmentary, but the pattern is consistent. Pythagoras did not begin with diagrams. He began with sound. Aristotle, writing in the fourth century BCE, reports that the Pythagoreans believed "things themselves are numbers" (*Metaphysics*, I.5). This was not an abstract slogan.

It emerged from observation. When different hammers struck metal in a smithy, some combinations rang true while others clashed. The behaviour was audible. The outcome was repeatable. Ratio, not force, determined harmony.

Listen first. That is the quiet motif threading this account. Pythagoras listened before he named. He attended to recurrence before he formalised law. The implication is precise. Mathematics, at its birth, was not an exercise in examination but an act of disciplined attention. The ear led the mind.

The smithy story itself can be traced to Nicomachus of Gerasa in the first century CE, who recounts that Pythagoras noticed consonance depended on the weight ratios of hammers. Modern historians note the physics is imperfect; pitch does not depend directly on hammer weight. The limitation is real. The anecdote cannot be taken as experimental fact. Yet dismissing it misses the point. The story records a way of noticing. It shows a thinker asking why some combinations please while others jar. The explanation is philosophical, not mechanical. Harmony points beyond itself.

That explanation mattered because it connected sensation to structure. A lyre string divided in half produces an octave. The behaviour is simple. Shorten the string. Pluck it. The metric is length. The outcome is consonance. The implication is that pleasure follows proportion. This relationship was teachable. It could be repeated by others. Music became a gateway to number.

The choice of music was not incidental. Sound occupies time. It unfolds. To notice pattern in sound is to notice order without freezing it. Pythagoras did not carve harmony into stone first; he let it pass through the body. The abstraction of ratio emerged from experience. Measure followed listening.

This order of operations stands in contrast to much later pedagogy. Modern mathematics education often begins with symbols and rules, postponing intuition. The Pythagorean approach reversed that sequence. Sensory pattern came first. Formalisation came later. The implication is not that ancient methods were superior, but that they preserved a relationship modern practice often severs.

The community that formed around Pythagoras at Croton treated number with seriousness that bordered on reverence. Initiates observed silence. They shared property. They studied music, astronomy, and geometry as aspects of a single inquiry. The behaviour was communal. The metric was harmony, not score. The implication was ethical. To live well required attunement, not accumulation.

One named practice illustrates this. According to Iamblichus, writing in the third century CE, Pythagoreans examined their actions each evening by recalling them in sequence. What did I do? Where did I fail? The behaviour involved counting moments. The abstraction was self-knowledge. The implication was accountability shaped by order. Number served reflection.

The sea returns as image. Waves do not strike at random. They arrive in sets, modulated by wind and tide. Anyone who has stood at the shore long enough notices recurrence without uniformity. Pythagoras, living near water, would have known this rhythm intimately. The motif holds. Listening reveals pattern without demanding control.

A second checkable example grounds this further. In Plato's *Timaeus*, composed around 360 BCE in Athens, the cosmos is described as structured by numerical ratios, especially those governing musical intervals. Plato names Pythagoras explicitly elsewhere and draws heavily on his school's ideas. The text is verifiable. The place and date are secure. The implication is enduring influence. Number, for Plato, mediates between the sensible and the intelligible.

This mediation matters. It prevents mathematics from collapsing into mere utility. For the Pythagoreans, number explained why the world was intelligible at all. The explanation was metaphysical, but it arose from concrete behaviours: listening, measuring, comparing. Trust in number grew because number disclosed order without erasing mystery.

Yet the approach carried risks. The same school that celebrated harmony also imposed secrecy. Certain ratios were treated as sacred. The famous crisis over irrational numbers, often associated with Hippasus of Metapontum in the fifth century BCE, illustrates the limitation. When it became clear that the diagonal of a square could not be expressed as a ratio of whole numbers, the tidy numerical cosmos fractured. Later sources claim Hippasus was drowned for revealing the discovery. The historicity of the punishment is disputed, but the shock is not. Number had failed to contain reality.

This counter-example matters because it exposes a fault line. When number is treated as complete, anomaly becomes threat. The implication is cautionary. Wonder curdles into dogma when openness closes. The Pythagorean legacy includes both insight and rigidity.

Reconciliation came slowly. Greek mathematics absorbed irrationality. Euclid's *Elements*, compiled around 300 BCE in Alexandria, systematised geometry without requiring all magnitudes to be commensurable. The behaviour shifted. Proof replaced proclamation. The implication was stability gained at the cost of some immediacy. Listening gave way to demonstration.

Even so, the original posture did not vanish. It retreated. Music remained part of mathematical education in the quadrivium of late antiquity and the medieval period. To study number was still, in part, to study harmony. The behaviour persisted in cloisters and courts, long after the myths hardened.

The relevance of this history lies not in antiquarian interest but in contrast. Modern professionals often encounter mathematics as an instrument of assessment. Examinations rank. Metrics decide. The behaviour is compliance. The abstraction is merit. The implication is pressure. Wonder has little room to breathe.

The Pythagorean scene offers an alternative starting point. Imagine a curriculum that begins not with tests but with tuned strings. The metric would be resonance. The behaviour would be comparison without competition. This is not a proposal. It is a diagnostic. Where does learning begin: with fear or with attention?

Listening appears again in contemporary science, though under different names. In signal processing, patterns are detected in noise by attending to frequency and amplitude. The behaviour is analysis. The metric is spectrum. The abstraction is information. The implication is that even the most advanced mathematics retains an auditory metaphor. Signals are "heard" against background.

This continuity does not collapse differences between ancient and modern practice. It clarifies lineage. Pythagoras listened for number because number revealed itself as relation. Modern mathematicians model relations with symbols. The gap is real. The bridge is attention.

The shoreline image recurs for a final time. Standing by the sea, one cannot command waves. One can only observe. That posture shaped an entire tradition. Mathematics emerged not as domination but as dialogue. When later centuries forgot this, number hardened into authority.

The unresolved tension remains. Can a culture that relies on numbers for decision and control recover a way of listening to them instead? Or has the distance grown too wide? The test is practical. When faced with a metric, do we ask only whether it is correct, or also what pattern it invites us to hear?

From tally marks to systems

In 1960, the Belgian geologist Jean de Heinzelin examined a bone recovered from Ishango, near Lake Edward on the border of present-day Democratic Republic of the Congo and Uganda. Radiocarbon dating later placed the artefact at roughly 20,000 BCE. The bone bore three columns of incised notches, carefully spaced and grouped. Heinzelin published his findings in Brussels the same year. The outcome was immediate debate. The marks were deliberate. Someone had counted. The place, the date, the object remain verifiable. The question has never settled.

The bone does not explain itself. It lies between tool and text. The notches are too regular to be accidental, too spare to be decorative. They form sequences. Groups repeat. Some appear to cluster around specific quantities. The behaviour implied is clear enough: a hand returned again and again to the same surface, committing marks in time. Count first. That is the quiet instruction the object seems to give.

The five-word sentence belongs here. Marks outlast memory.

This is the starting point for any serious account of number. Before symbols, before names, before systems, there was inscription. Counting began when memory needed help. The explanation is practical. A hunter tracking days between migrations, a gatherer monitoring yield, a group coordinating effort. The implication is sobering. Number emerged not from leisure, but from risk.

Show the act. A notch is cut. Another follows. The spacing matters. Each cut commits a moment to permanence. Name the claim. Tally marks externalise trust. Explain it. Once quantity is marked, it no longer depends on recollection or reputation alone. Others can verify it. The implication follows directly. Counting makes cooperation possible beyond immediate presence.

The Ishango bone is not unique. Similar tally artefacts have been found across Europe, Africa, and Asia, dated to the Upper Palaeolithic period. The Abri Blanchard bone from France, dated to around 30,000 BCE, carries grouped incisions that many archaeologists interpret as lunar counts. Alexander Marshack argued this case in the 1970s, suggesting the marks tracked phases of the moon. His interpretation has been challenged, but the limitation sharpens rather than weakens the point. Whether lunar or not, the marks encode sequence. They insist on order.

The motif of the line returns here, cut into bone. A line remembers when bodies cannot. That line, once trusted, invites extension. Counting does not stop at the notch. It accumulates.

As groups grew larger and movement increased, tallying alone strained. Trade introduced delay. Goods changed hands without immediate return. Obligation stretched across time. The behaviour shifted. Marks needed to be portable, legible to strangers, resistant to dispute. This pressure drove the emergence of more formal systems.

By the late fourth millennium BCE, in Mesopotamia, clay tokens represented quantities of goods. Denise Schmandt-Besserat documented this system extensively in the 1970s and 1980s. Small shapes stood for jars of oil, measures of grain, livestock. The place is known. Uruk. The period is established. Around 3300 BCE, these tokens were pressed into clay tablets, creating impressions that became proto-cuneiform numerals.

The outcome was revolutionary. Quantity could now travel without the object.

Show the behaviour. A merchant seals a tablet. The marks correspond to tokens enclosed within. Name the claim. Numeral systems arise from administration. Explain it. When exchange scales beyond face-to-face trust, representation must stabilise meaning. The implication is structural. Bureaucracy and number co-evolve.

This is not abstraction for its own sake. The metric was concrete. Measures of barley. Counts of sheep. Days of labour owed. Efficiency increased because disputes could be resolved by reference to a shared record. Risk decreased because obligation was fixed. Counting, here, is a technology of peace.

Yet the cost was distance. The person who made the mark was not always the person who felt its consequence. The line, once intimate, became institutional. This distance would later enable symbolism, but it also introduced alienation. Numbers began to speak with an authority no individual could easily challenge.

Calendars developed under similar pressures. Agricultural societies required coordination across seasons. In ancient Egypt, by the third millennium BCE, a civil calendar of 365 days regulated taxation and labour. The heliacal rising of Sirius marked the Nile flood. The behaviour was observational. Priests watched the sky. The metric was days. The implication was survival. Plant too early or too late and crops failed.

The calendar embodies a critical shift. Counting time binds human action to natural cycles. It translates observation into obligation. Work begins here. Taxation follows there. The line now stretches across months and years. The motif persists.

A second named example reinforces the point. In Shang dynasty China, around 1200 BCE, oracle bones recorded divinations alongside date notations using a sexagenary cycle. The place is Anyang. The material is turtle plastron and ox scapula. The behaviour combined ritual and record. The metric was days within a cyclical system. The implication is striking. Even sacred enquiry relied on structured counting.

This convergence of ritual and record undermines a common misconception. Symbolism did not precede system. It emerged alongside it. Numbers acquired meaning because they already carried consequence. Seven mattered because days mattered. Twelve mattered because months mattered. Later interpretations rested on earlier necessities.

The limitation must be faced. Not all cultures developed formal numeral systems. Some hunter-gatherer groups operated with limited counting words, sufficient for immediate needs. The Pirahã people of the Amazon, documented by Daniel Everett in the early 2000s, famously use approximate quantity terms rather than exact numerals. The behaviour is observable. Precision is not prioritised. The implication is clear. Extensive systems are not inevitable.

Reconciliation lies in context. Where exchange remains local and immediate, approximation suffices. Where obligation extends across distance and time, exactitude pays. Systems respond to scale. They do not float free of it.

As systems matured, they gained internal elegance. The Babylonian base-60 system, attested by the early second millennium BCE, enabled sophisticated astronomy. Sixty divides cleanly. The metric facilitated calculation of angles and time. The implication endures. We still measure hours and circles in its shadow. Efficiency, here, is literal. Calculations shorten.

The abstraction is often celebrated. The behaviour that sustained it is less remarked. Scribes trained for years. Errors carried penalties. Trust in the system depended on disciplined bodies. Counting at scale demands labour.

The line motif surfaces again in columns of numerals. Vertical alignment replaces horizontal scratch. The orientation shifts, but the function remains. Order holds.

By the time Greek mathematicians encountered these systems, number had already accumulated authority. When later symbolism framed number as cosmic principle, it drew legitimacy from this long record of reliability. People trusted numbers because numbers had fed them.

This grounding matters for the broader argument of the book. Sacred mathematics does not descend from speculation alone. It rises from repeated success. When counting works, it earns reverence. When it fails, trust erodes.

The unresolved tension is present even here. Systems that stabilise cooperation also enable control. Calendars organise labour, but they also fix obligation. Records settle disputes, but they also surveil. The same line that once remembered for the group can later constrain it.

The forward test remains open. When we encounter a numerical system, do we ask what scale of survival it once served, and what scale of power it now enforces? The tally mark has not vanished. It has multiplied. Whether it still remembers for us, or now remembers us, is the question that follows.

Fear of numbers, hunger for certainty

In May 2013, in Indianapolis, the mathematician and science communicator John Allen Paulos wrote in *The New York Times* about a familiar spectacle: intelligent adults turning away from numbers with a mixture of embarrassment and irritation, as if quantity were a foreign language they were tired of pretending to speak. Paulos had been documenting this phenomenon since his 1988 book *Innumeracy*, and the outcome he described was not academic. People misread risk, misunderstand debt, and make avoidable errors in judgement because they cannot bear to look at figures for long. The place and date are checkable. The behaviour is ordinary. The consequences are not.

A household bill arrives. A line of digits sits beside a currency symbol. The sum is not enormous, but the reaction is immediate: a quick inhale, a tightening of the shoulders, the urge to postpone. Another envelope contains a medical report, or a health news item: a percentage, a ratio, a risk figure. The mind searches for a verdict. Am I safe? Am I ruined? Quantity appears to offer a clean answer, then refuses to behave like one.

Numbers reassure and threaten in the same instant. That is the paradox.

The question is not why numbers matter. The question is why they sting. Adults who can read complex prose, negotiate contracts, and manage teams can still feel childish dread when faced with basic figures. This is not mere incompetence. It is a relationship problem. It involves memory, shame, and the cultural authority attached to measurement.

Show the behaviour first. A professional scans an invoice and looks away. They open a banking app and wince at a red notification. They read a headline containing a percentage and interpret it as certainty rather than probability. Name the claim. Numerical anxiety is widespread and consequential. Explain it with evidence rather than folklore. Psychologists have studied what they call mathematics anxiety for decades; a frequently cited paper by Mark H. Ashcraft and Elizabeth P. Kirk (2001) describes how anxiety consumes working memory, reducing performance during calculation. The implication is practical. Fear impairs competence, and impaired competence feeds fear. The cycle tightens.

This matters because modern life routes decisions through quantified interfaces. Credit scores influence loan approvals. Performance targets determine promotion. Risk estimates inform screening and treatment decisions. The abstraction of "security" becomes a visible behaviour: checking the number, then obeying it. People do not merely consult quantities; they submit to them, often without knowing what they measure or how they were derived.

Bills provide a clean example because they combine arithmetic with emotion. A bill is not only a sum. It is a story about obligation. It arrives with authority and a deadline. The metric is due date. The decision is pay now, delay, dispute, or ignore. A numerate response requires parsing: unit price, usage, interest, penalties. But parsing requires time, and time is scarce. The result is predictable: many people treat the bill as a verdict rather than as information, and the response becomes reactive.

In the United Kingdom, the introduction of smart meters was partly sold as a way to make energy consumption legible in real time. The promise was visible: a display showing kilowatt-hours and estimated cost. Legibility, however, is not the same as comfort. A number that updates constantly can soothe some households and unsettle others. The behaviour shifts from paying a quarterly bill to monitoring daily consumption. The metric becomes immediate. The implication is psychological. When numbers move closer, they can either empower or harry.

Medical statistics sharpen the stakes. Here, the hunger for certainty becomes almost desperate. A risk figure seems to answer the oldest questions: How long? How likely? How serious? Yet statistics rarely speak in the language the frightened mind wants. Relative risk and absolute risk can tell different stories. A risk reduction of "50%" can mean a fall from two cases in a thousand to one in a thousand, or from twenty in a hundred to ten in a hundred. The behaviour that follows depends on which interpretation a person carries into the consultation room or the late-night search.

This is not a niche concern. In 2007, the physician and researcher Gerd Gigerenzer, based at the Max Planck Institute for Human Development in Berlin, argued in *BMJ* that both patients and doctors often struggle with basic risk literacy, leading to misunderstandings about screening and treatment. He proposed using "natural frequencies" (such as "1 in 10" rather than "10%") to improve comprehension. The place and year are checkable. The claim is testable. The implication is concrete. How we present numbers changes decisions.

The argument can be misread as a complaint about intelligence. It is not. It is about format, context, and trust. A frightened person does not process information like a calm one. A busy professional does not read percentages with the patience of a statistician. When numbers appear without narrative scaffolding, the mind supplies its own, often with catastrophic flair.

Trust is often invoked here as an abstraction, so it deserves a visible anchor. Consider the decision to accept a screening test. The metric might be sensitivity, specificity, false positive rate. The behaviour is consent. The implication is downstream: anxiety, follow-up procedures, reassurance, or delay. When these metrics are not understood, trust shifts from informed assent to deference. Deference is not the same as trust. It is submission.

Fear, however, is not the whole story. Adults also crave numbers. They reach for them the way a hand reaches for a railing in the dark. The banking app, the credit score, the blood pressure reading, the productivity dashboard. The comfort comes from having something to hold. A number seems stable, and stability feels like safety. Even people who "hate maths" often track something. Steps. Weight. Hours worked. Likes. These are acts of counting performed in private, where the risk of judgement feels lower.

This is where the hunger for certainty becomes culturally consequential. Numbers offer the appearance of objectivity. They promise a world that can be managed. Yet the promise is partial, and the disappointment can be cruel. The desire for certainty drives over-measurement, and over-measurement drives more anxiety when the measured reality fails to behave. A person sets a savings target, misses it, and feels moral failure rather than financial inconvenience. A researcher watches citation counts and mistakes them for intellectual worth. A student equates a grade with identity.

The motif that threads this section is the ledger. Not merely the accounting ledger, but the mental one: the quiet book we keep of credits and debits, of "enough" and "not enough". The ledger appears in bills, in bank balances, in medical probabilities, in performance rankings. It structures attention. It encourages comparison. It can also devour the imagination.

Show the ledger at work. A manager reviews a quarterly report and fixates on a single number: churn, revenue, defect rate. The decision becomes simple: cut costs, pressure staff, change strategy. The explanation is not that metrics are wrong, but that they narrow attention. The implication is ethical. When a metric becomes the whole story, human judgement shrinks.

This narrowing is not accidental. Sociologists and economists have long warned of the perverse incentives created by measurement regimes. Charles Goodhart, a British economist, is often associated with the observation later popularised as "when a measure becomes a target, it ceases to be a good measure" (Goodhart's Law), discussed in policy contexts from the 1970s onward. The behaviour is visible: people optimise for the metric. The metric is the target. The implication is distortion. Counting shapes behaviour, whether or not the count captures what matters.

A credible counter-example must be admitted. There are contexts where numerical precision saves lives and reduces harm. Aviation provides a clean case. Checklists, weight limits, fuel calculations, maintenance intervals: these are not optional. The behaviour is strict compliance. The metric is safety margins. The implication is survival. Numbers here are not enemies; they are guardians. The fear some feel when confronted with such figures does not make the figures less necessary.

Reconciliation requires discernment. The issue is not numbers versus meaning. It is which numbers, for what purpose, and with what humility. In aviation, metrics are audited, continually revised, and embedded in a culture that treats error as information. In many personal and organisational contexts, metrics arrive without that culture. They are treated as verdicts rather than tools. The same digit can function as guidance or as judgement depending on the surrounding practice.

This helps explain why adults flinch. Many people have experienced numbers in settings where error carried humiliation. The classroom. The workplace review. The public scoreboard. Shame teaches avoidance quickly. Once avoidance becomes habit, the mind loses fluency. The next encounter becomes harder. The ledger grows heavier.

The antidote is not forced optimism. It is curiosity with boundaries. Curiosity asks: what does this number *actually* measure? How was it produced? What does it omit? A bill can be read as a composition of units rather than as a moral accusation. A medical statistic can be held as an estimate rather than as a sentence. These are not therapeutic instructions; they are epistemic habits, ways of relating to information without collapsing into it.

A number is never alone. It always implies a frame. This is where reverence begins to re-enter the discussion. Reverence is not superstition. It is respect for the power of abstraction. Numbers can compress vast realities into a few symbols. That compression is astonishing. It deserves seriousness. But seriousness differs from submission. Reverence can include scepticism. It can include refusal.

The ledger motif returns here as a diagnostic. When you see a number, do you read it as a line in a ledger of worth, or as a line in a ledger of information? The difference is behavioural. In the first case, you avert your eyes or obsess. In the second, you compare, check, and decide.

This subchapter sets up a move that will matter for the rest of the book: from treating numbers as enemies or idols to treating them as interlocutors. Bills and medical statistics show the stakes because they touch survival, status, and fear. The question is not whether numbers will govern us. They already do. The question is whether we will meet them with a flinch, or with a steadier attention that can tolerate uncertainty without demanding that a digit relieve it.

The forward test is simple to state and hard to live. When the next number arrives—your balance, your risk figure, your target—do you rush to obey it, rush to deny it, or pause long enough to ask what kind of certainty it is selling?

Chapter 2
The Harmony of Ratios

A guitar string is plucked, then pressed lightly at its midpoint and plucked again. The sound changes, yet nothing mysterious has happened. The string is the same. The hands are the same. Only the length has altered. What emerges is an octave—clean, recognisable, almost inevitable. No calculation is required to feel that the second note *fits* the first. The body knows it before the mind names it. Ratio has announced itself through the ear.

This small experiment unsettles a deeply ingrained assumption. Numbers, we are often told, are abstractions imposed on the world, tools we use to tame unruly phenomena. Yet here, number does not flatten experience; it sharpens it. The fraction one-half is not an idea hovering above sound. It *is* the sound. Ratio becomes audible, tangible, trustworthy. Mathematics steps out of the page and into the air.

That movement—from symbol to sensation—is the threshold this chapter crosses. If the previous chapter traced how counting establishes order, this one asks a subtler question: how does proportion create harmony? Not harmony as metaphor alone, but as something felt in the body, recognised across cultures, and quietly relied upon in daily life. Ratios do not merely count; they relate. They bind difference into coherence.

Music provides the most direct entry point because it bypasses argument. A perfect fifth does not need defending. It persuades by resonance. Long before ratios were written as fractions, they were heard as intervals. Strings, pipes, voices. The simplicity of the relationships—two to one, three to two—invited trust because they could be tested repeatedly with the senses. Anyone could listen. Anyone could feel when something was out of tune.

This sensory foundation matters because it explains why ratios acquired such philosophical weight. They were not invented to decorate theory. They were recognised because they worked. When ancient thinkers spoke of harmony, they did not mean mere pleasantness. They meant lawful relationship. Difference held without collapse. Tension resolved without erasure. These are not musical concerns alone; they are existential ones.

The chapter follows this thread outward, from instruments to the cosmos. The so-called music of the spheres is often dismissed as a quaint error, a relic of pre-scientific imagination. Planets, we now know, do not emit audible chords as they circle the sun. Yet to treat the idea as foolish is to miss its ambition. It was an attempt to articulate a conviction that the universe is ordered, that motion follows ratio even when the ear cannot hear it. Number here carried truth and imagination together, not in competition but in partnership.

That partnership remains fragile. Modern science rightly insists on accuracy, measurement, and testable claims. Poetic cosmology cannot substitute for orbital mechanics. Yet something is lost when imagination is banished entirely. The ancient image of a resonant cosmos was not a technical model; it was a statement of belonging. It suggested that human beings, by recognising harmony in sound, were participating in a larger order rather than observing it from outside.

This sense of participation reappears when ratio is traced into the body and the built environment. Proportion is not confined to instruments or heavens. It shapes rooms, streets, and faces. The difference between a cramped space and a generous one is rarely reducible to square footage alone. Ceiling height, window placement, the relation between width and length—all contribute to how a body settles or tenses. These relationships are numerical, but they are apprehended somatically. One does not need to know the ratios to feel their effects.

Classical architecture understood this intimately. Temples and civic buildings were designed so that movement through them unfolded in measured sequences. Steps, columns, intervals. The body learned proportion by walking. Yet the same principle operates in humbler settings: a kitchen laid out so that reaching, turning, and resting follow a rhythm; a staircase whose rise and run invite ascent rather than resistance. Ratio, here, is not ornament. It is hospitality.

Art and anatomy offer similar lessons. The human eye is remarkably sensitive to proportion. Faces that deviate too far from certain relational patterns register as unsettling, even when individual features are unremarkable. This is not a matter of rigid templates or idealised formulas; it is a matter of balance. The body, like music, tolerates variation within limits. Ratio defines those limits without fixing them absolutely.

The chapter does not argue for a single canon of proportion. That path leads quickly to dogma. Instead, it attends to experience. What happens when proportions work? Ease. What happens when they do not? Fatigue, irritation, disorientation. These responses are data, even if they resist easy quantification. They remind us that ratio is not an abstract ideal imposed from above, but a lived relationship negotiated between form and perception.

From here, the narrative turns toward a historical shift with lasting consequences. Over time, the musical and architectural sense of ratio hardened into precise measurement. Fractions became decimals. Proportions became specifications. This transformation enabled extraordinary achievements. Bridges span rivers because loads are calculated. Recipes scale because quantities are fixed. Precision saves lives and resources. The gain is undeniable.

Yet something subtle changed in the process. Measurement began to overshadow harmony. The question shifted from "does this fit?" to "does this meet the specification?" Often the answers coincide. Sometimes they do not. A structure can meet every requirement and still feel wrong. A process can be efficient and still exhaust those who follow it. When ratio loses its sensory grounding, it risks becoming brittle.

This chapter does not romanticise a return to vagueness. Precision matters. But it asks whether precision can remain in dialogue with proportion rather than replacing it. The example of a recipe is telling. Cooking by weight ensures repeatability. Cooking by eye and taste ensures responsiveness. Skilled cooks do both. They measure, then adjust. Harmony lives in that adjustment.

Modern relevance presses in here with urgency. Digital systems optimise relentlessly. Algorithms tune ratios for engagement, throughput, conversion. These are forms of harmony, but of a narrow kind. When a feed feels addictive rather than nourishing, when a workflow maximises output at the cost of attention, the issue is not number itself but which ratios are privileged. The ancient question returns in a new guise: what relationships are we honouring?

A quiet motif threads the chapter: tuning. Instruments are tuned before they are played. Rooms are tuned by proportion. Bodies tune themselves through posture and movement. Even organisations speak of being "out of tune" when processes clash. Tuning implies care, listening, adjustment. It resists finality. A tuned instrument will drift and must be tuned again. Harmony is not a state achieved once and for all; it is a practice.

The chapter's scope is deliberately wide, but its movement is controlled. It begins with sound, expands to cosmos, settles into body and building, and then tightens into the practicalities of measurement. Each step preserves continuity with what came before. Counting established order. Ratio establishes relationship. Together, they form a grammar that can be spoken crudely or with fluency.

The emotional payoff promised is not aesthetic alone. It is ethical. When ratios are honoured, systems tend to respect limits. When they are ignored, systems overreach. The harmony of ratios teaches restraint as much as creativity. It suggests that more is not always better, and that balance is not the enemy of progress.

As the chapter unfolds, readers will be invited to listen differently—to sounds, to spaces, to the quiet signals of fit and strain that numbers alone cannot capture. The invitation is not to abandon measurement, but to re-embed it in experience. To let number be heard as well as seen.

The pages ahead do not offer formulas to memorise. They offer a way of noticing. If counting taught us to line things up, ratio teaches us how tightly or loosely they should be held together. The difference matters. The world hums with relationships. Whether we experience them as noise or as music depends, in part, on whether we have learned how to tune.

Music as visible number

In 1636, in Paris, the Minim friar and polymath Marin Mersenne published *Harmonie universelle*, a vast treatise that set out, with uncommon patience, the numerical foundations of musical sound. Mersenne was not speculating at a distance. He stretched strings, weighed them, shortened them, and listened. The outcome was precise. A string half the length of another produced a pitch an octave higher. A ratio of three to two yielded a perfect fifth. These relationships could be demonstrated repeatedly, on lutes, viols, and monochords, before any sceptical ear. Mersenne did not ask readers to believe. He asked them to hear.

The significance of that request is easy to underestimate. Music offered something rare: number that announced itself through sensation. Ratio stepped out of abstraction and into air. A fraction ceased to be a mark on a page and became a felt difference in the chest, a tightening or easing of the ear. Mathematics, in this form, did not require prior trust. It generated trust by working.

This matters because it explains why music has long served as a gateway to numerical understanding. People who recoil from symbols will still hum an octave. They will still notice when two notes sit well together, and when they do not. The body recognises proportionality without instruction. Ratio arrives as experience.

Hear it first.

The core claim can be shown simply. Take a familiar instrument. A guitar will do. Pluck an open string. Then press lightly at its midpoint and pluck again. The second sound is not random. It is the octave, produced because the vibrating length has been halved. Show the act. The hand divides the string. Name the relationship. Two to one. Explain the mechanism. Frequency doubles as length halves. The implication follows. A numerical relationship has become audible, and its stability invites repetition.

This invitation is crucial. Trust grows through repeatability. When an experiment works every time, confidence shifts from authority to experience. Music trained generations of listeners to accept number not because a teacher declared it true, but because their ears confirmed it. That habit of confirmation matters far beyond sound.

Mersenne's work stands as a checkable example, but the insight predates him by millennia. Pythagorean musicians in the sixth and fifth centuries BCE used the monochord to demonstrate ratios to students. A single string stretched over a resonant box, marked at measured intervals, allowed pitch to be altered by moving a bridge. The behaviour was demonstrative. The metric was length. The implication was pedagogical. Anyone could see and hear the same result.

The monochord embodies the chapter's theme. It makes number visible without reducing it to notation. The string becomes a line that can be divided, marked, and sounded. The motif returns. Lines hold order.

This sensory grounding explains why musical ratios retained philosophical authority even as other numerical claims faltered. When ancient thinkers argued about geometry or astronomy, disagreement was common. When they demonstrated consonance, agreement followed. The ear settled disputes faster than the eye. Harmony persuaded.

Name the abstraction carefully. "Harmony" here does not mean prettiness. It means fit. The Greek term *harmonia* referred to joining, fastening, bringing parts into relation. In music, that joining is temporal. Notes follow one another, overlap, resolve. The behaviour is audible. The metric is interval. The implication extends. Ratio becomes a model for how difference can coexist without collapse.

This model travelled widely. In medieval Europe, music formed one quarter of the quadrivium, alongside arithmetic, geometry, and astronomy. Students learned proportions through chant before encountering algebraic notation. The behaviour was communal singing. The metric was interval and rhythm. The implication was cognitive. Ratio was learned in the body before it was formalised on the page.

A second named example sharpens the point. In 1722, in Leipzig, Johann Sebastian Bach compiled *Das wohltemperierte Klavier*, a collection of keyboard pieces designed to explore all twenty-four major and minor keys. The work presupposed a tuning system that balanced pure ratios against practical compromise. Equal temperament slightly adjusted intervals so that modulation across keys became possible. The decision was deliberate. Bach accepted small numerical impurities to gain expressive range. The implication is instructive. Harmony can tolerate distortion when guided by purpose.

This introduces a necessary limitation. Musical ratios are not absolute. Perfect fifths sound clean because they align closely with simple fractions, but tuning systems differ across cultures and eras. The equal-tempered scale used in most Western music today distributes error evenly across intervals. Purity is sacrificed for flexibility. The behaviour is tuning. The metric is cents, a logarithmic measure of pitch difference. The implication is nuanced. Mathematical relationships can be adapted without losing their guiding role.

This adaptation matters because it prevents idealisation. Music does not offer a frozen model of numerical truth. It offers a practice of negotiation. Ratios guide, but they do not dictate. Musicians listen, adjust, and commit. That posture resists the tyranny of exactness without abandoning structure.

Trust reappears here as a visible behaviour. An orchestra tunes before playing. Each musician adjusts pitch relative to a reference tone. The metric is alignment. The decision is when to stop adjusting and begin. The implication is communal. Harmony is achieved together, not individually.

This communal dimension distinguishes musical number from many other applications of mathematics. In accounting or engineering, numbers often adjudicate. In music, they coordinate. A singer who drifts sharp is not immoral; they are out of tune. The correction is immediate and shared. The body responds before judgement forms.

Contrast this with a classroom exam. A wrong answer produces a mark. The metric is score. The implication is rank.

The emotional tone shifts. Music teaches number without humiliation. That difference is not trivial. It shapes long-term attitudes.

Modern research supports this experiential distinction. Studies in cognitive science have shown that rhythmic and melodic training can support numerical reasoning, particularly in early education. While the specifics vary, the pattern holds: when number is linked to embodied pattern, engagement increases. The behaviour is tapping, singing, moving. The metric is timing and pitch. The implication is pedagogical, though this book resists prescription.

Music also clarifies risk. A wrong note does not destroy the piece. It resolves or it passes. The cost is low. The system encourages experimentation. Mathematics, when stripped of such low-stakes exploration, becomes brittle. Fear follows.

The motif of tuning returns. Instruments drift with temperature and time. They must be retuned. Ratio is not a possession; it is a maintenance task. This impermanence matters. It reminds us that numerical relationships live in conditions. They respond to context.

Modern technologies often obscure this responsiveness. Digital synthesisers generate pitches with mathematical exactness. The metric is frequency to several decimal places. The sound is stable. Yet many musicians still prefer analogue instruments because their slight instabilities produce warmth. The behaviour is choice. The implication is aesthetic. Precision alone does not guarantee satisfaction.

This does not undermine mathematics. It situates it. Number describes possibilities; judgement selects among them. Music makes that selection audible.

The abstraction of "enjoyment" deserves a concrete anchor. Consider the simple act of singing in unison. When voices align, vibration synchronises. The metric is frequency match. The behaviour is listening and adjusting. The implication is physiological. Breathing slows. Attention steadies. Trust grows because the body feels held by order rather than threatened by it.

This bodily trust is what allows music to carry number across cultures. Scales differ. Instruments differ. Ratios recur. The octave appears in traditions separated by geography and language because halving a vibrating length produces the same result everywhere. The universality is not ideological. It is mechanical. Yet the meaning drawn from it varies.

Here lies the unresolved tension that closes the subchapter. Music makes number sensible and humane, but it also reveals limits. Ratios that delight the ear do not map neatly onto every domain. The danger is extrapolation without care. The promise is integration without reduction.

The forward test is simple. When encountering a ratio—whether in sound, space, or system—do we treat it as a command to obey, or as an invitation to listen? The difference will determine whether mathematics remains an ally of attention, or hardens once again into a language spoken only under duress.

The music of the spheres

In 1619, in Linz, Johannes Kepler published *Harmonices Mundi* (*The Harmony of the World*), a book written with the intensity of a man who believed the heavens could be understood not only as machinery but as meaning. Kepler was Imperial Mathematician to the Holy Roman Emperor, a role that demanded calculation with unforgiving precision, yet he chose the language of music. He compared planetary motions to intervals. He searched for consonance in their changing speeds. He did not merely compute orbits; he listened for structure within them, convinced that lawful motion could be expressed as harmony.

Kepler's work produced a concrete outcome. In *Harmonices Mundi* he stated what became known as his third law: a mathematical relationship linking a planet's orbital period to its distance from the sun. The law is testable. It remains a foundation of celestial mechanics. Yet Kepler framed this durable result inside an imaginative ambition: that the cosmos, properly understood, is a form of ordered resonance.

He aimed to hear order.

The phrase "music of the spheres" is older than Kepler, older even than the texts that repeat it most elegantly. It is a shorthand for a proposal: the universe is not a chaos of arbitrary events but a patterned whole, and number is the grammar of that pattern. The phrase misleads if taken literally. Planets do not generate audible chords as they travel. Space is not filled with air to carry their sound. The idea fails as acoustics. Its endurance lies elsewhere. It is a way of holding together two human needs that do not always coexist peacefully: accuracy and imagination.

Show the need. A sailor looks up and navigates by stars. An administrator sets a calendar and collects tax by season. A scholar measures angles and predicts eclipses. These are practical acts. Name the claim. Celestial regularity trains trust in law. Explain it. When the sky behaves predictably, the mind expects order elsewhere. The implication is philosophical. A lawful cosmos invites meaning without requiring superstition.

The music of the spheres was one of the earliest attempts to articulate that invitation. It took a sensory certainty—musical consonance—and used it as a model for a non-sensory domain—the heavens. Ratios that pleased the ear became a template for thinking about motion, distance, and time. This was not a childish mistake; it was a disciplined analogy. Like all analogies, it risked overreach. Like the best analogies, it created a space where thought could move.

The most influential early articulation sits in a verifiable place and text. In Athens around 360 BCE, Plato composed the *Timaeus*, a dialogue that describes the world as a crafted order, structured by mathematical ratios. Plato's cosmology is not a physics textbook. It is a philosophical portrait. Yet it makes a decisive move: it binds cosmic order to number, and it implies that the human mind can recognise that order because it shares in the same rational structure. The behaviour implied is intellectual confidence. The metric is proportion. The implication is bold. Understanding the heavens becomes a form of self-recognition.

To modern readers, the temptation is to sort these ideas into two piles: wrong astronomy, right philosophy. That sorting is too crude. The ancient claim did not arise from ignorance of observation; it arose from a different relationship to observation. The sky was not merely "out there" as an object. It was a communal clock, a source of orientation, a boundary that demanded interpretation. When Greeks and later thinkers spoke of harmony among planetary motions, they were not only proposing mechanisms; they were making a wager about intelligibility.

A second named example reveals the wager's persistence under the pressure of evidence. Kepler's era inherited a world in which astronomical models had become increasingly quantitative. Tycho Brahe's meticulous observations, recorded at Uraniborg on the island of Hven in the late sixteenth century, offered a new standard for accuracy. Kepler, working with Brahe's data after 1601, faced discrepancies that could not be smoothed away with elegance alone. Mars in particular refused to behave. Kepler's response was not to abandon harmony, but to refine what harmony could mean. He committed to calculation so strict that it broke old models. He also refused to surrender the conviction that lawful structure was beautiful.

Show the behaviour. Kepler recalculates. He rejects circular orbits. He accepts ellipses because the numbers force the concession. Name the claim. Harmony survives when it submits to evidence. Explain it. If harmony is treated as a verdict, anomalies become threats; if treated as a hypothesis, anomalies become instruction. The implication is methodological. Wonder becomes durable when it partners with correction.

This partnership is where the music of the spheres earns respect rather than indulgence. Kepler's musical language did not prevent him from making hard, falsifiable claims. It drove him to them. He wanted the cosmos to be coherent enough to sing. When the data demanded a new tune, he changed it.

The limitation, however, remains essential. Literal harmony misrepresents the nature of planetary motion. Ratios in orbit are not fixed like ratios on a monochord. A planet speeds up and slows down along its path. The "intervals" Kepler sought were not stable notes; they were changing relationships. The metaphor strains. It can seduce the mind into hearing music where there is only pattern. This is the moment where a poetic image, if clung to too tightly, becomes an obstacle.

Name the counter-example plainly. A lawful universe does not guarantee a musical one. Explain it without disdain. Musical consonance depends on specific physical conditions: vibrating bodies, a medium, and ears built to interpret pressure changes as pitch. Planetary motion produces no such direct experience. The implication is intellectual humility. Not every pattern is a song.

Reconciliation comes from recognising what the metaphor was doing, and doing it with clearer rules. The music of the spheres was never merely about sound. It was about commensurability: the hope that complex motion could be expressed in relationships simple enough to be grasped, tested, and communicated. Even when the metaphor fails acoustically, it can still succeed epistemically. It invites the mind to search for relationships that remain stable under transformation.

That epistemic success becomes visible when astronomy turns into mechanics. In 1687, in London, Isaac Newton published *Philosophiæ Naturalis Principia Mathematica*. Newton did not write in musical terms. He wrote in the language of force, mass, and geometry. Yet the intellectual gesture is strangely similar to the ancient one: a small set of mathematical principles is proposed as sufficient to describe vast celestial behaviour. Newton's achievement tightened the bond between number and nature so firmly that it became hard to imagine them apart.

Show the behaviour. Newton uses Kepler's laws as evidence and derives them from universal gravitation. Name the claim. Mathematical law can compress the cosmos. Explain it. When one equation accounts for both falling apples and planetary orbits, the mind learns to trust abstraction. The implication is cultural. Numbers gain authority because they predict.

This is where the music of the spheres becomes more than historical curiosity. It marks a recurring tension in the human response to lawful nature. Prediction comforts. It also tempts. Once numbers predict, we begin to treat them as final rather than provisional. The same hunger for certainty that leads adults to cling to medical statistics can lead cultures to cling to cosmic laws as if they were complete. The music of the spheres, properly understood, is a warning as much as an inspiration: it reminds us that lawfulness can coexist with mystery.

A quiet motif threads through this section: resonance. In music, resonance is the amplification of a note by a body that shares its frequency. In thought, resonance is the sense that an idea "fits" the world. The Pythagorean and Platonic traditions treated cosmic order as resonant with human reason. Kepler treated it as resonant with both reason and aesthetic necessity. Newton treated it as resonant with mathematical structure. Each approach made a different decision about what kind of resonance mattered.

The decisions had consequences. When resonance is defined too narrowly, the world feels reduced. When defined too loosely, the world becomes a mirror for our wishes. The challenge is to hold resonance as a disciplined feeling: a sense of fit that must be tested and revised.

Modern readers live with the fruits and the costs of this history. We inhabit a universe described with extraordinary mathematical precision, yet we also live amid narratives that treat numbers as cold, hostile, and spiritually empty. The ancient image of cosmic music offers a way to interrupt that story. It does not ask us to believe in literal planetary choirs. It asks us to notice that mathematical law can be experienced not only as control, but as coherence.

Coherence is not comfort. It can be unsettling. A lawful universe implies constraints. Planets do not choose their paths. Seasons do not negotiate. If we read lawfulness only as determinism, we may recoil. If we read it as reliability, we may relax. The same pattern can be experienced as imprisonment or as shelter, depending on our posture towards it.

This posture shapes practical behaviour. Consider the difference between prediction and understanding. A model may forecast an eclipse with minute accuracy, yet tell us little about the felt meaning of standing under a darkened sun. The metric is precision. The decision is how to interpret it. The implication is human. Numbers can describe events without exhausting experience.

That gap is not a failure of science. It is a reminder of domain. The music of the spheres tried to bridge the gap by borrowing the language of sensation. It sought to hold together the predictable and the profound. We still need that bridge. We just need it built with clearer materials.

This is why the chapter balances image and accuracy rather than sacrificing one to the other. Poetic cosmology without correction becomes fantasy. Technical description without imagination becomes mere inventory. The ancient and early modern tradition of cosmic harmony offers a third option: disciplined wonder. It is wonder that expects to be corrected, and correction that does not extinguish wonder.

The forward test is neither theological nor technical. It is interpretive. When a numerical law describes the heavens with chilling elegance, do we experience that elegance as a closure—proof that everything important has been said—or as an opening, a sign that intelligibility itself is a mystery worth contemplating?

Proportion in body and building

In 1490, in Milan, Leonardo da Vinci completed a pen-and-ink drawing that would become one of the most reproduced images in Western art: the *Vitruvian Man*. The sheet shows a naked male figure inscribed within a circle and a square, arms and legs extended to mark precise relationships between span, height, and centre. Leonardo based the drawing on the first-century BCE Roman architect Vitruvius, whose treatise *De Architectura* argued that buildings should follow the proportions of the human body. The outcome was not decorative. It was diagnostic. Architecture, Vitruvius claimed, succeeds when it resonates with embodied measure.

The drawing still unsettles because it reverses an assumption. Proportion does not begin in walls and columns; it begins in flesh. The body is not accommodated by good design as an afterthought. It is the template. Long before anyone named ratios explicitly, they were felt as reach, balance, strain, and ease. Space was judged by how it received a body moving through it.

You feel proportion immediately.

Enter a low-ceilinged corridor and shoulders tighten. Step into a room with generous height and breath deepens. The behaviour is involuntary. The metric is posture and respiration. The implication is direct. Ratio operates beneath awareness, shaping experience before judgement forms. This is not aesthetic theory; it is physiology.

Show the contrast. A cramped stairwell forces the body into caution. Steps feel steeper than they are. A well-proportioned staircase invites ascent. The rise and run align with gait. The decision to climb feels effortless. Name the claim. Proportion guides movement. Explain it. When spatial ratios align with bodily rhythms, effort decreases. The implication is practical. Good proportion conserves energy.

This conservation extends beyond comfort. It shapes trust. In spaces that fit, bodies relax. In spaces that fight the body, vigilance increases. Architects have long known this, even when they did not articulate it numerically. Medieval builders adjusted nave heights and aisle widths through experience, refining proportions until processions flowed without congestion and sound carried without distortion. The metric was not a formula alone; it was use.

Leonardo's drawing formalised an intuition that had guided building for centuries. Vitruvius specified that the span of outstretched arms equals height, that the navel marks a centre. These claims are not universally precise. Bodies vary. The limitation matters. The *Vitruvian Man* is an idealisation, not a census. Yet the drawing persists because it captures a relational truth: the body is a system of ratios. Design that ignores those ratios courts resistance.

A second named example grounds this further. In 1951, in Marseille, the architect Le Corbusier completed the *Unité d'Habitation*, a massive residential block based on a proportional system he called the Modulor. The Modulor combined human measurements with the golden ratio, producing a scale intended to guide everything from ceiling heights to door handles. The place and date are fixed. The outcome was mixed. Some residents reported ease and clarity; others felt oppressed by the rigidity of the system.

The mixed reception is instructive. Show the behaviour. Residents live, work, and rest within a proportional grid. Name the claim. Proportion alone does not guarantee hospitality. Explain it. When ratios are applied inflexibly, variation suffers. The implication is cautionary. Sacred proportion becomes tyrannical when it forgets contingency.

This counter-example matters because it prevents myth-making. There is no single ratio that sanctifies a space. Proportion is relational, not magical. It must answer to context, culture, and use. The body that Vitruvius described was a Roman male citizen's body. The Modulor assumed a particular height and stride. When bodies differ, the ratios must adapt.

Reconciliation comes through practice rather than doctrine. Skilled builders adjust. They test. They watch how people move. Proportion emerges through feedback. The behaviour is iterative. The metric is occupancy and flow. The implication is humane. Ratio serves life when it listens.

The body itself offers the clearest evidence. Anatomy is not symmetrical perfection. It is functional proportion. Limbs balance torque. Organs fit cavities. The ribcage expands to admit breath. These relationships are not optimised for beauty but for survival. Yet beauty often follows. When proportion aligns with function, form satisfies.

Artists have long exploited this alignment. Classical sculpture exaggerates or suppresses certain ratios to convey strength, grace, or calm. The *Doryphoros* of Polykleitos, created in the fifth century BCE, was accompanied by a lost treatise on proportion known as the *Canon*. Polykleitos argued that beauty arises from precise relationships between parts. The statue stands in contrapposto, weight shifted, balance achieved through counter-tension. The behaviour is visual scanning. The metric is relative size. The implication is perceptual. The eye recognises coherence.

Yet even here, rigidity fails. Later Hellenistic sculpture stretched proportions to convey drama. Baroque architecture bent classical ratios to create movement. Bodies and buildings both resist being frozen. Proportion must flex to express life.

Return to domestic scale. A kitchen where counters sit too low strains the back. One where they sit too high strains the shoulders. The metric is centimetres. The behaviour is repeated reach. The implication is cumulative fatigue or ease. This is not abstract. It is felt daily.

The same holds for furniture. Chairs designed without regard to thigh length or lumbar curve tire the sitter. Chairs that respect bodily proportion invite rest. The decision to remain seated becomes effortless. Trust builds quietly.

This trust extends to sacred spaces. Cathedrals, mosques, temples—whatever their doctrinal differences—often share an attention to proportion that regulates movement and gaze. A nave widens gradually. Light enters at calculated heights. Sound decays at a rate that supports chant or prayer. The behaviour is walking, standing, listening. The metric is reverberation time, step cadence, sightline. The implication is contemplative. Bodies are guided without instruction.

Modern research has begun to name these effects. Studies in environmental psychology link ceiling height to cognitive processing styles, suggesting that higher ceilings encourage abstract thinking while lower ceilings promote focus. The claim is probabilistic, not absolute. The behaviour measured is task performance. The implication is modest but telling. Spatial ratios influence thought.

This influence complicates the claim that proportion is merely aesthetic. It shapes cognition. It frames interaction. It can calm or agitate. When spaces ignore this, occupants compensate with effort. When spaces attend to it, occupants conserve attention.

The motif threading this section is the doorway. A doorway mediates between spaces. Its width and height signal what kind of passage is expected. Duck, pause, stride. The body reads the ratio instantly. Doorways teach proportion through action.

Doorways also symbolise thresholds in the argument. Proportion sits between abstraction and sensation. It is number experienced as passage. Too narrow, and movement stalls. Too wide, and orientation dissolves. Fit matters.

The contemporary built environment often struggles here. Economic pressures favour standardisation. Ratios are optimised for cost, not experience. The metric is square metres per unit. The decision is density. The implication is visible. Many new dwellings feel tight even when technically adequate. Bodies adapt, but uneasily.

This is not a call for extravagance. It is a call for attention. Proportion need not increase cost dramatically; it requires care. Slight changes in ceiling height or window placement can transform experience. The difference lies in whether ratios are treated as afterthoughts or as organising principles.

Art, anatomy, and architecture converge on a shared lesson. Proportion is not an ornament applied once function is solved. It *is* a function. It governs how bodies inhabit form. When honoured, it disappears into ease. When ignored, it announces itself as discomfort.

The unresolved tension lies here. Modern tools allow us to calculate proportions with extraordinary precision. Yet precision alone does not guarantee fit. The decision remains human. Which ratios deserve allegiance? Which bodies count as reference? Which forms will be asked to bend?

The forward test is lived, not theoretical. Pay attention to the next space you enter. Does your body settle or brace? The answer arrives before thought. Proportion has already spoken.

From harmony to measurement

In May 1843, in Bristol, Isambard Kingdom Brunel opened the Clifton Suspension Bridge to the public—at least in the sense that the structure, long imagined and contested, finally began to take its recognisable form over the Avon Gorge. The bridge as seen today was completed later, but its governing idea belongs to Brunel's era: a span held by calculated tension and compression, daring in silhouette yet anchored in arithmetic. The outcome was not merely a crossing. It was a statement that number could hold beauty aloft without asking it to apologise for itself.

The bridge is a useful place to begin because it makes measurement visible. You can stand beneath its cables and feel the risk that has been disciplined into reliability. You can also feel the grace. The line of the deck is not an accident. Neither is its survival. The same ratios that once tuned strings now guide loads, stresses, and safety margins. Harmony did not vanish; it hardened.

Five words belong here. Precision carries weight and hope.

The story of that hardening is not a tale of progress replacing superstition. It is a tale of translation: the movement from sensed proportion to quantified specification, from "this fits" to "this will hold". The shift began slowly and unevenly. Artisans measured with bodies and tools long before they wrote equations. Cathedrals rose without calculus. Bread was baked without grams. Yet across centuries, a change took place. Ratio, once primarily heard and felt, became something that could be standardised, replicated, audited.

Show the earlier posture. A stonemason judges an arch by eye and experience. The behaviour is iterative adjustment. The metric is fit and stability. The implication is local mastery. The knowledge lives in hands. Now name the later posture. An engineer calculates load paths before building. The behaviour is modelling. The metric is stress, deflection, factor of safety. The implication is portability. Knowledge moves from person to paper.

This portability altered civilisation. It allowed complex structures to be designed away from the site, components to be manufactured elsewhere, and failure to be analysed after the fact. But it also introduced a risk of its own: the risk that numbers would be treated as sufficient, that the living sense of proportion would be reduced to compliance with specification. The best engineering avoids that reduction. The worst invites it.

The bridge offers a disciplined example of both forces. Clifton's eventual completion relied on more than Brunel's eye. It relied on calculations, materials testing, and an emerging culture of civil engineering in Britain. Yet its enduring appeal is aesthetic. People photograph it not because it is safe, but because it is right. The curve of cable, the spacing of towers, the relation between span and height—these are proportional choices that resonate beyond utility.

Name the claim. Measurement can serve beauty. Explain it. Beauty in structures often arises when constraints are resolved elegantly rather than merely met. The implication is cultural. When engineering forgets this, the built environment becomes merely adequate.

To see how ratio hardened, it helps to follow measurement into the kitchen. Cooking is a domestic laboratory where proportion is both strict and forgiving. A recipe for bread depends on ratios: flour to water, yeast to salt. Bakers speak of hydration as a percentage. The behaviour is weighing and mixing. The metric is grams and temperature. The implication is reproducibility. Bread rises similarly across kitchens because ratio governs transformation.

Yet any competent cook knows the limitation. A recipe cannot fully account for humidity, flour type, yeast vigour, or the temperament of an oven. The baker reads the dough. They adjust water by feel. They lengthen fermentation because the room is cold. The decision is not arbitrary; it is informed by experience. The implication is precise. Measurement alone does not guarantee success. Proportion must remain embodied.

This embodied adjustment echoes the musical practice of tuning. A score specifies notes, but performers shape timbre and timing. A recipe specifies quantities, but cooks respond to conditions. A bridge design specifies tolerances, but builders respond to site. The motif that threads this piece is the *calibration*—the moment when abstract quantity meets material reality and demands judgement.

Calibration is the hinge between harmony and measurement. It is also where trust is won or lost. A measurement system that calibrates well becomes reliable. One that ignores context produces brittle failures.

A named example of measurement's rise in Europe clarifies the stakes. In 1795, in revolutionary France, the metre was defined as a fraction of the Earth's meridian, and the metric system began its slow expansion. The project was political as well as scientific: a bid to replace local measures with a universal standard. The behaviour was administrative: retooling commerce, signage, manufacture. The metric was literally the metre and kilogram. The implication was profound. Standardisation enabled exchange at scale.

Standardisation also changed craft. When components can be specified in standard units, production can be distributed. A bridge's bolts can be made in one place and shipped to another. A recipe can be multiplied in an industrial kitchen. Efficiency increases. The decisions become quicker because ambiguity decreases.

Yet ambiguity has its uses. Local measures often encoded local realities: a "cup" in one region suited local ingredients and vessels. When standardisation arrives, it can flatten the relationship between practice and place. This is not a reason to reject standards; it is a reminder that standards carry assumptions.

The counter-example here is failure. When measurement becomes detached from material reality, structures fail dramatically. The history of engineering is littered with such lessons. A famous modern instance is the Mars Climate Orbiter, lost in 1999 because of a mismatch between imperial and metric units in software calculations, as reported by NASA. The behaviour was a unit conversion error. The metric was thrust calculation. The implication was costly. Precision without shared conventions collapses into nonsense.

This counter-example is not offered to mock. It clarifies a principle. Measurement is a social system as much as a numerical one. It depends on shared definitions, disciplined communication, and humility about error. Without those, the most refined calculation becomes a trap.

Reconciliation arrives when we treat measurement as an extension of harmony rather than its replacement. Harmony begins as sensed fit. Measurement formalises that fit so it can be communicated, tested, and improved. The bridge and the recipe both show this. In each case, beauty and utility are not competitors. They are intertwined outcomes of ratio applied with judgement.

This intertwining becomes clearer when we consider what engineers call a factor of safety. The factor is a numerical cushion, the ratio between what a structure should withstand and what it is expected to face. The behaviour is conservative design. The metric is often two, three, or more, depending on context. The implication is ethical. Engineers build humility into numbers. They acknowledge uncertainty.

This is one of measurement's quiet spiritual lessons, whether or not one uses religious language. Good measurement does not claim omniscience. It quantifies risk and then steps back from the edge. It assumes error and plans for it. The best ratios include mercy.

Cooking has its own version of this. A recipe rarely demands exactness to the decimal; it allows tolerance. Salt "to taste" is not vagueness; it is calibration. The cook uses a sensory test. The metric is tongue and timing. The implication is quality. Taste, like structural integrity, cannot be reduced entirely to specification.

The relationship between beauty and utility also appears in the design of everyday objects. Consider a well-made chair. Its dimensions must accommodate bodies. Its joints must bear load. Yet its proportions also invite rest. Good furniture designers, from the Shakers to modernists, have long understood that the elegance of an object arises when function is met with minimal strain and maximal clarity. The behaviour is sitting. The metric is angle, height, weight distribution. The implication is lived. A chair teaches ratio by receiving the body.

The danger arises when measurement becomes the only language permitted. Organisations then treat metrics as proxies for value. Engineers may be pressured to optimise for cost at the expense of durability. Kitchens may be pressured to standardise flavour at the expense of craft. Buildings may be designed to meet regulations while ignoring experience. The metric becomes compliance. The behaviour becomes minimal satisfaction of requirements. The implication is a world that works, yet feels wrong.

The answer is not to abandon measurement. It is to remember that measurement came from harmony. The earliest ratios were trusted because they were felt. The best modern measurements remain faithful to that origin by staying close to experience and by making room for judgement. When a bridge is both safe and beautiful, when a recipe is both repeatable and alive, we witness that fidelity.

Calibration returns as motif. A musician calibrates pitch. A cook calibrates dough. An engineer calibrates models against tests and failures. Calibration is the practice of humility that keeps numbers honest.

The unresolved tension is this: standardisation enables scale, but scale tempts simplification. The more we rely on measurement, the more we risk forgetting what measurement is for. The forward test is not technical. It is interpretive. When faced with a specification, a metric, a quantified target, do we ask only whether it is met, or whether it produces a form of harmony that bodies can live inside?

Chapter 3
The Grammar of Quantity

A loaf of bread sits on the table. There are four people and five slices. The problem is not arithmetic, yet arithmetic is immediately summoned. Who eats less? Who waits? Do we cut a slice in half, or accept an imbalance? In that small, ordinary moment, numbers step forward not as abstractions but as participants in a moral scene. Quantity does not merely describe the situation; it shapes the possible responses. Counting, here, is already a language of choice.

We rarely notice how fluently we speak this language until it falters. A bank balance slips below zero and suddenly a minus sign acquires weight. A cupboard is empty and zero announces itself, not as a neutral symbol but as absence. A bill must be split and fractions appear, not as school exercises but as negotiations between people. Numbers arrive with tone and force. They behave less like inert marks and more like words that can console, accuse, or clarify.

This chapter treats numbers in that spirit: not as a single undifferentiated tool, but as a family with distinct roles. Natural numbers, zero, negatives, fractions, infinities—each enters human life through a different door and does different work once inside. Together they form a grammar of quantity, a way of structuring experience that is as subtle, and as consequential, as spoken language. We do not merely *use* numbers; we speak with them, often without noticing the assumptions embedded in their syntax.

The idea that mathematics functions like a language is not new, but it is often misunderstood. Language is not just vocabulary. It has parts of speech, rules of combination, and unspoken conventions about emphasis and omission. So does number. Counting "one, two, three" performs a different act from saying "none". Writing "–£50" carries a different emotional charge from writing "£50". A fraction implies sharing; an integer often implies possession. These nuances are learned through life long before they are formalised.

The chapter begins by slowing this process down. Natural numbers are the nouns of the numerical world: discrete, countable, reassuring. They suit objects you can point to and line up. Zero enters as a grammatical shock, a word that names nothing yet changes everything. It allows us to mark absence, to keep place, to distinguish between "not yet" and "never". Negatives arrive later, often through debt or obligation, reversing the direction of quantity and forcing us to imagine less than nothing. Fractions mediate between wholes, teaching compromise and proportion. Each of these "characters" appears first in lived scenes, not on blackboards.

Seen this way, numeracy is not merely competence; it is fluency. Some people struggle with mathematics not because they lack intelligence, but because they were never taught to hear its grammar. They memorised rules without grasping tone. They learned syntax without context. Reframing numbers as parts of speech allows a different approach, one grounded in meaning rather than fear.

From this family gathering, the chapter moves toward a more vertiginous edge. Infinity appears, not as a technical object to be manipulated, but as a boundary concept that tests the limits of language itself. A traveller walks and never reaches the end. Two mirrors face each other and produce a corridor of reflections without last image. Infinity resists counting even as it invites it. It is not a number one can hold; it is a horizon that recedes as one approaches.

Encounters with infinity often provoke cleverness—paradoxes, proofs, intellectual games. This chapter insists on a different centre of gravity. Awe belongs here. Infinity matters not because it sharpens wit, but because it exposes finitude. It reminds us that any grammar of quantity, however refined, operates within limits. The human mind can gesture beyond those limits, but cannot abolish them. That gesture has spiritual weight precisely because it fails to close.

The presence of infinity raises an old and persistent question: are numbers discovered or invented? Do they exist independently of us, waiting to be found, or are they grids we lay over experience to make it manageable? The chapter approaches this debate without technical armour, using domestic acts as entry points. When you measure a table, are you uncovering a property of the wood, or imposing a unit of length chosen by convention? When you count stars, are you revealing a cosmic truth, or merely exhausting your patience?

The question matters because it shapes posture. If numbers are discovered, they command a certain reverence. They are part of the fabric of reality, and learning them is a form of listening. If they are invented, they demand responsibility. They reflect our choices, priorities, and blind spots. In practice, most serious thinkers have occupied a space between these extremes, recognising that number emerges from an encounter between mind and world. This chapter prepares the ground for that encounter by keeping the examples homely and the stakes clear.

Those stakes become explicit when the discussion turns from what numbers *are* to what we *count*. Counting is never neutral. To count something is to declare it worthy of attention. To leave something uncounted is not necessarily to dismiss it, but it is to risk forgetting it. Modern life overflows with metrics, dashboards, and rankings that claim to make value visible. Yet many of the things people care about most—trust, dignity, love, wisdom—resist easy quantification.

This tension is not an abstract problem. It plays out daily in workplaces, schools, and homes. Productivity is measured; care is assumed. Screen time is tracked; boredom is not. Steps are counted; rest is improvised. The chapter does not argue against measurement, but it insists on discernment. What we count shapes what we become attentive to, and what we neglect shapes us just as surely.

A subtle motif runs through these explorations: the act of naming. To name a number is to stabilise a relationship. To refuse to name is sometimes an act of humility, sometimes of avoidance. Language teaches us that silence can be meaningful. So can uncounted space. The grammar of quantity includes not only what is spoken, but what is left unsaid.

The continuity with earlier chapters is deliberate. Counting established order. Ratio revealed relationship. Grammar now introduces meaning and nuance. The book's argument deepens here, moving from external pattern to internal interpretation. Numbers no longer appear merely as structures in the world, but as elements in a conversation between human judgement and reality's resistance.

Modern relevance presses in quietly. Artificial intelligence systems count relentlessly. Social platforms quantify attention. Economic models reduce complexity to variables that can be optimised. These systems are powerful precisely because they speak a simplified grammar fluently. The danger is not that they count, but that they count without listening for what falls outside their syntax. This chapter does not preach caution; it equips readers to recognise when grammar has been mistaken for truth.

By the time the chapter turns its gaze forward, the reader will have encountered numbers not as a monolith, but as a varied and expressive language. Natural numbers, zero, negatives, fractions, infinity—each will have taken its place in a wider conversation about meaning, limit, and choice. The philosophical questions raised will not be resolved here. They are not meant to be. They are meant to sharpen attention.

The closing movement does not offer answers; it offers orientation. If numbers are a language, then fluency involves more than calculation. It involves knowing when to speak, when to pause, and when to admit that no available word quite fits. The chapters that follow will return to this insight in different guises—symbolic, geometric, scientific, devotional—but the grammar is established here.

What remains is an invitation. The next time a number appears in your life, ask not only whether it is correct, but what role it is playing. Is it naming, dividing, erasing, or extending? Is it closing a conversation or opening one? Grammar always shapes thought. Quantity is no exception.

The family of numbers

In around 825 CE, in Baghdad, the scholar Muhammad ibn Mūsā al-Khwarizmi completed a treatise that quietly changed how quantities could speak. Working in the House of Wisdom, he set out methods for calculation that treated *sifr*—the empty place—as a legitimate participant in reckoning. The outcome was practical. Accounts balanced. Procedures simplified. What had once been an awkward gap became a usable symbol. Zero entered the family not as an idea, but as a working member.

Absence gained a voice.

That moment matters because it exposes a common misunderstanding. Numbers do not arrive as a single, uniform block. They arrive over time, each carrying a distinct role. Counting stones requires one kind of number. Marking debt requires another. Sharing food introduces a third. Naming what is missing demands a fourth. Treating mathematics as a monolith flattens these differences and invites confusion. Treating it as a language restores them.

The language metaphor earns its keep here. Languages have nouns, verbs, and modifiers. They have tense and negation. They distinguish between presence and absence, possession and obligation. So do numbers. Natural numbers count what is there. Zero marks what is not. Negative numbers reverse direction. Fractions negotiate division without destruction. Each does work the others cannot. Together, they form a grammar that structures daily life long before it appears in formal study.

Begin with the simplest members. Natural numbers enter first, almost unnoticed. One apple. Two chairs. Three days. They behave like nouns, naming discrete things that can be pointed to and gathered. The behaviour is lining up. The metric is count. The implication is order. Natural numbers reassure because they stabilise. They tell us that the world can be enumerated without dissolving.

This reassurance is visible in childhood play and adult logistics alike. When a delivery arrives with five boxes, counting confirms completeness. When attendance is taken, numbers confirm presence. Trust grows because the count can be checked. Natural numbers are conservative. They conserve identity by keeping items separate.

Yet their limits appear quickly. A cupboard can be empty. A seat can be unoccupied. A promise can fail to arrive. Without a way to name nothing, counting stalls. This is where zero steps forward.

Zero behaves less like a noun and more like punctuation. It marks place. It distinguishes "no apples" from "forgot to count". In positional notation, its role is decisive. Without it, 205 collapses into 25. The behaviour is spacing. The metric is position. The implication is clarity. Zero allows structure to scale.

Al-Khwarizmi's contribution was not philosophical musing. It was procedural clarity. By treating zero as something that could be written, moved, and operated upon, he enabled calculations that would otherwise tangle. The empty place became productive. Absence became legible.

This legibility carries emotional weight. A balance of zero can relieve anxiety. A score of zero can sting. The symbol itself is neutral; its context is not. Zero marks a threshold between possession and lack. It allows systems to pause without collapsing.

The family grows more complex when quantity moves below zero. Negative numbers often arrive through discomfort. Debt is their most common teacher. A bank statement showing -£50 does not describe an abstract inversion; it describes obligation. The behaviour is owing. The metric is deficit. The implication is urgency.

Historically, negative numbers were resisted for centuries. They felt unnatural because they named what could not be held. In China, by the first century CE, *The Nine Chapters on the Mathematical Art* used red and black counting rods to represent positive and negative quantities in solving commercial problems. The place and text are verifiable. The behaviour was calculation. The implication was pragmatic. Trade demanded a way to represent loss as well as gain.

Europe took longer to accept this move. For medieval accountants, debt was often handled rhetorically rather than numerically. Only with the expansion of commerce did negatives become unavoidable. Once accepted, they transformed accounting and algebra. Direction entered number. Quantity acquired orientation.

Orientation matters. Natural numbers accumulate. Negatives point. They indicate movement away from a baseline. Without them, many economic and physical descriptions remain incomplete. Temperature below freezing, elevation below sea level, financial liability—all require a sign that reverses expectation.

The discomfort they provoke persists. People will often say they have "less than nothing" with a grim smile. The phrase is mathematically precise. The feeling is not. Negative numbers remind us that grammar shapes emotion. A minus sign alters tone.

Fractions introduce yet another role. Where natural numbers separate, fractions bind. They arise in scenes of sharing. A loaf divided among four people. A bill split between friends. The behaviour is cutting or allocating. The metric is proportion. The implication is fairness.

Fractions resist the all-or-nothing logic that integers encourage. They acknowledge that wholes can be preserved while portions change. This makes them socially potent. A half implies compromise. A quarter implies restraint. Fractions teach negotiation by necessity.

Their historical development reflects this. In ancient Egypt, fractions were written as sums of unit fractions, emphasising sharing into equal parts. The Rhind Mathematical Papyrus, dated to around 1650 BCE, records problems about distributing bread and beer. The place and date are fixed. The behaviour is division. The implication is social coordination.

Fractions also unsettle. They complicate counting by introducing endlessness into what seemed finite. One divided by three does not terminate cleanly. Decimal expansions repeat. The family shows its strain. Not everything fits neatly.

This strain reveals a limitation that deserves attention. Treating numbers as characters risks anthropomorphism. Numbers do not have intentions. They do not care. The metaphor works only if it illuminates function without mystifying agency. The reconciliation lies in discipline. Characters here are roles, not beings. They describe how different quantities behave in practice.

Seen this way, the family metaphor clarifies why numerical confusion often arises. People fluent with natural numbers may still struggle with fractions. Those comfortable with counting may recoil from negatives. The difficulty is grammatical. It resembles mistaking an adjective for a noun. The rules differ.

Consider trust as a visible behaviour. A person trusts a financial summary that shows income, expenditure, and balance because the grammar is complete. Natural numbers list transactions. Negatives show debts. Zero marks equilibrium. Fractions might appear in interest rates. Omit one element and trust erodes. The metric becomes misleading.

Efficiency offers another example. A recipe written only in whole numbers wastes material. "Add one cup" works until the scale changes. Fractions allow scaling without distortion. The behaviour is adjustment. The implication is economy. Grammar enables efficiency.

Risk becomes legible through negatives and fractions. A probability of one in ten does not command the same response as a loss of ten pounds. The difference lies in grammatical role. One describes chance; the other describes direction. Confusing them leads to error.

The motif threading this subchapter is the ledger. Not the spreadsheet, but the act of keeping account. A ledger requires multiple numerical roles. Credits and debits. Balances and fractions. The ledger fails if any role is excluded. Life does the same.

Modern systems often simplify this grammar for speed. Dashboards favour totals over distributions. Scores replace narratives. The behaviour is aggregation. The metric is headline figures. The implication is blindness to nuance. When grammar is reduced, meaning thins.

The family of numbers resists that reduction by insisting on diversity of function. Natural numbers count. Zero holds place.

Negatives reverse. Fractions relate. None can replace the others without loss.

The unresolved tension appears when systems privilege one role excessively. When only totals matter, absence is ignored. When only deficits matter, growth is discounted. When only fractions matter, wholes dissolve. Grammar requires balance.

The forward test is practical. When you encounter a number, ask which family member is speaking and what role it is playing. Is it naming, negating, dividing, or orienting? Confusion often begins when we answer the wrong question.

Infinity at the edge of thought

In 1874, in Halle, the mathematician Georg Cantor published a paper with the dry title "Über eine Eigenschaft des Inbegriffes aller reellen algebraischen Zahlen" and detonated a quiet revolution. Cantor proved that the set of real numbers cannot be listed in the same way as the set of natural numbers. There are, in his terms, different sizes of infinity. The outcome was not merely a technical result. It triggered a century of philosophical turbulence and professional dispute, including public criticism from Leopold Kronecker in Berlin, who rejected Cantor's infinities as metaphysical excess. The place and year are checkable. The argument remains.

Cantor's proof is often treated as a clever trick, a magician's flourish in a lecture theatre. Yet the most honest response is not applause. It is vertigo. Infinity, once approached, rearranges the furniture of thought. It does not sit politely in the mind like a well-behaved number. It presses against the boundaries of language itself and reveals where our usual grammar breaks.

Infinity refuses completion.

That refusal is what makes it spiritually significant without requiring religious claims. Infinity forces the mind to confront a limit: the limit of counting, the limit of imagining, the limit of possession. A person can count stones until tiredness intervenes; they cannot count the natural numbers to the end because there is no end. A traveller can walk for years; they cannot walk to the edge of an unbounded line. Infinity is not a destination. It is the recognition that some destinations cannot exist.

Show the traveller. A figure walks along a road that stretches beyond the visible horizon. The behaviour is steady pacing. The metric is distance measured in steps or miles. The decision is whether to continue. The implication is existential. Every finite segment can be completed, yet the whole cannot. The road offers progress without arrival.

This is the first lesson infinity teaches. It distinguishes between the finite and the inexhaustible. In the finite, effort can in principle finish its object. In the inexhaustible, effort can only continue. The distinction is not merely mathematical; it governs how humans relate to work, knowledge, and desire. When people treat inexhaustible tasks as if they were finite—when they expect to "finish" understanding, finish achievement, finish certainty—they invite frustration. Infinity reveals why.

The second lesson is subtler. Infinity can be approached in more than one way. Sometimes it appears as endless addition: one more, and then one more. Sometimes it appears as endless division: half, then half again, and again. These two gestures carry different emotional tones. The first suggests accumulation. The second suggests refinement. Both resist closure.

The ancient Greeks were already haunted by these gestures. Zeno of Elea, writing in the fifth century BCE, offered paradoxes that exposed the uneasy relationship between motion and infinite division. Achilles, he argued, cannot catch a tortoise if he must first reach the point where the tortoise began, then the next point, and so on without end. The story is verifiable in Aristotle's *Physics*. The behaviour is pursuit. The metric is distance subdivided. The implication is perplexity: how can motion occur if it requires completing infinitely many steps?

Modern readers often treat Zeno as a clever nuisance, a man who mistook the map for the road. Yet Zeno's paradoxes remain instructive because they force a separation between mathematical description and lived experience. We do walk. Achilles does catch the tortoise in practice. But the mind, when it describes the path as infinitely divisible, stumbles. Infinity here is not a cosmic quantity; it is a feature of our representational systems. The paradox exposes a boundary, not a mistake.

This boundary appears in a more domestic form: two mirrors facing each other. Stand between them and a corridor of reflections opens, each image smaller and dimmer, receding without visible end. The behaviour is looking. The metric is repetition, a count of images that cannot be completed because it depends on resolution and light. The implication is visceral. Infinity can be *seen* as repetition without termination, even though one knows the corridor is only a play of light.

The mirror corridor is a useful motif because it unites the chapter's purpose. Infinity is not a thing you can hold; it is a pattern you can recognise. The corridor is real enough to unsettle and insubstantial enough to resist capture. That is infinity's characteristic status in thought: compelling, elusive, and relentlessly instructive.

Cantor's work brought this elusive pattern into sharp technical focus. Before him, many mathematicians treated infinity as a limit concept, something gestured towards but never treated as an object. Cantor insisted that infinite sets could be compared. He distinguished between countable infinity—the infinity of the natural numbers—and the uncountable infinity of the real numbers. The behaviour is definition and proof. The metric is cardinality. The implication is startling. Infinity is not one uniform vastness; it has structure.

This is where many readers reach for cleverness as a defence. If infinity is made into a puzzle, it becomes manageable. But cleverness can function as avoidance. Awe is the more appropriate centre. Cantor's result does not merely offer a new trick; it undermines the assumption that the mind's intuition about "more" maps neatly onto reality. When infinities differ in size, "more" becomes a fragile notion.

Trust, as an abstraction, must be anchored in behaviour here. Trust in mathematics means accepting results that run against intuition because the argument compels. The behaviour is assent to proof. The metric is logical validity. The implication is discipline: the mind must submit, not to authority, but to coherence.

Yet this discipline has limits. Cantor's programme did not resolve all questions about infinity; it opened new ones. At the beginning of the twentieth century, paradoxes in set theory—such as Russell's paradox, published by Bertrand Russell in 1901—showed that naïve talk about "the set of all sets" generates contradictions. The behaviour is unrestricted definition. The metric is consistency. The implication is crisis. Infinity, handled carelessly, breaks systems.

This is the credible counter-example. Infinity does not merely stretch thought; it can tear it. When people treat infinite totalities as if they were ordinary objects, contradictions emerge. The reconciliation was not to ban infinity, but to develop stricter foundations. David Hilbert's programme and later axiomatic set theories attempted to specify the rules under which infinity could be safely discussed. The behaviour is formalisation. The metric is consistency proofs and axioms. The implication is humility. Infinity requires guardrails.

The spiritual parallel is precise without being preachy. Some concepts demand reverence because they exceed our usual modes of handling. Infinity is one of them. Reverence, here, means disciplined approach and admitted limits. It means refusing to confuse the symbol for the thing. It means accepting that language can point beyond itself without capturing what it points to.

The traveller motif returns. Imagine walking towards a mountain that recedes because the land is curved, or because fog thickens, or because the path spirals. The behaviour is continuing. The metric is time spent. The implication is diagnostic. Does the traveller walk with resentment at the lack of arrival, or with patience that treats the walking itself as meaningful? Infinity forces this question.

Modern professionals encounter analogous infinities daily, though they rarely name them. An inbox never empties for long. A news feed never ends. A research agenda expands with every answer. The behaviour is scrolling, checking, reading. The metric is unread counts, time-on-task, citations. The implication is fatigue when these are treated as finite tasks. Infinity here is manufactured, but the psychological effect resembles the mathematical boundary: continuation without completion.

This is not an argument against systems. It is an invitation to discernment. When a task is inherently unending, the correct metric is not "completion" but "enough". Yet "enough" is not a number. It is a judgement. Infinity forces judgement back into view.

The mirror corridor returns for a final time. Each reflection is real as an image and unreal as a thing. The corridor extends as far as light and resolution allow. It has no last picture you can point to. Infinity, likewise, is an image that deepens as attention deepens. You can keep looking. You will not reach the end.

The forward test is practical and uncomfortable. When you face something that feels infinite—knowledge, obligation, desire—do you reach for a false ending, or do you set a boundary that acknowledges the endlessness without surrendering to it? Infinity will not answer. It will only reveal how you respond.

Do we discover or invent numbers?

In 1802, in Göttingen, Carl Friedrich Gauss began publishing work that would earn him a title he did not request: *princeps mathematicorum*, the prince of mathematicians. Among his early results was a proof concerning the construction of a regular 17-gon with straightedge and compass, announced in 1796 and elaborated in later writings. The outcome was tangible. A shape thought impossible within classical constraints turned out to be possible. Gauss's discovery did not depend on taste or convention. It held whether or not anyone admired it. For many, that is the decisive evidence: mathematics is found, not made.

And yet the same Gauss, working as a surveyor and astronomer, also helped refine geodetic measurement by choosing coordinate systems and reference points. The Earth did not hand him a grid. He imposed one to make land measurable. Both gestures—discovering an unexpected polygon and laying a coordinate net over a landscape—occur in the same life. That tension is not a philosophical sideshow. It is the problem itself.

Numbers feel inevitable. They are also optional.

Consider a homely scene. A table sits in a room. You measure it to see whether it will fit through a doorway. The behaviour is mundane. You extend a tape measure, note the length, compare it with the door's width. The metric is centimetres or inches. The decision is purchase or refusal, moving the table or leaving it behind. The table's size seems like a fact about the table. Yet your measurement depends on choices: which unit, where you start, how tightly you hold the tape, whether you include the skirting board. The table does not contain "180 cm" inside it like a hidden label. You produced that number through an interaction.

Show the second scene. You stand outside at night and count stars. The behaviour is pointing and tallying. The metric is count. The decision is when to stop, because attention and skyglow limit what can be seen. The stars are there independently of you, but "how many" depends on your aperture—your eyes, your location, your patience, your category. Do you count the faint smudges at the edge of vision? Do you count a double star as one or two? The act of counting is a claim about what counts as an object.

The philosophical debate over whether numbers are discovered or invented begins precisely here, in the everyday friction between inevitability and choice. Treat it as a question about intellectual history alone and it becomes a parlour sport. Treat it as a question about how we relate to reality and it becomes ethically and spiritually charged. If numbers are discovered, they invite reverence: we are reading a script written into the world. If numbers are invented, they invite responsibility: we are writing a script that will shape behaviour, allocate resources, and justify decisions.

To hold this tension with integrity, we need to name the positions clearly without turning them into caricatures. One family of views—often labelled Platonism after Plato—treats numbers as real entities existing independently of human minds. On this view, "two" is not a human convention; it is a feature of reality that would persist even if no one counted.

Mathematical truths are discovered in the same sense that geographical facts are discovered: they are already the case.

Another family of views—often associated with formalism or conventionalism—treats mathematics as a human-made system of symbols governed by rules. On this view, numbers are not found in nature; they are defined in languages of our own making. Mathematical truth becomes a matter of consistency within a chosen system, like the truth of moves in chess.

Both positions sound clean until tested against actual practice. Mathematics behaves like discovery when it surprises us with results we did not expect, cannot change, and can only accept. It behaves like invention when it depends on definitions, notations, and axioms we choose. The debate persists because the evidence points both ways.

Show discovery at work. A mathematician proves a theorem and finds a result that feels unavoidable. The behaviour is logical derivation. The metric is proof. The implication is constraint: once the premises are fixed, the conclusion follows whether you like it or not. Gauss's 17-gon sits here. So does Cantor's hierarchy of infinities. So do the prime numbers, stubbornly patterned and stubbornly resistant to simple description.

Now show invention at work. A committee agrees on units. In 1983, the General Conference on Weights and Measures (CGPM) defined the metre in terms of the speed of light: the distance light travels in vacuum in 1/299,792,458 of a second. The organisation, year, and definition are checkable. The behaviour is standard-setting. The metric becomes enforceable. The implication is social: measurement depends on agreement. The universe did not decree "metre"; we did.

The two gestures—proof and convention—live in the same scientific culture. Engineers calculate using definitions that could have been otherwise, yet the results they predict either match reality or fail. When a bridge holds, it is not because the units were spiritually correct; it is because the relationships modelled were. The bridge teaches a reconciliation: the symbols may be invented, but the constraints they track are discovered.

This reconciliation can be named more precisely. Mathematics is an invented language that discovers real structure. The language—symbols, axioms, units—is chosen. The structure—relations, invariants, constraints—is encountered. The implication is that neither camp, taken alone, describes the full phenomenon. Treat mathematics as pure discovery and you underplay human creativity. Treat it as pure invention and you underplay the world's resistance.

The world resists in measurable ways. A table will not pass through a door if the width exceeds the opening, regardless of whether you measure in centimetres or inches. The behaviour of scraping wood against a frame supplies a kind of proof. The metric is fit. The implication is realism: some constraints do not negotiate.

Yet our grids matter. The doorframe can be considered an obstacle or a feature depending on the project. We decide what to measure based on what we value. In architecture, the same space can be described by area, volume, or accessibility metrics. Each count produces different priorities. This is why the discover-or-invent question matters spiritually. It shapes humility.

Humility appears as behaviour. A scientist revises a model when data disagree. A surveyor adjusts a coordinate system when error accumulates. A patient asks whether a risk statistic is absolute or relative. In each case, the number is not a decree; it is a provisional representation. The metric is discrepancy between prediction and observation. The implication is moral: obedience to numbers without judgement becomes superstition with better typography.

This is where the debate intersects with the hunger for certainty explored earlier. People often treat numbers as if they were discovered truths when it suits their desire for firmness, and treat them as flexible inventions when it suits their desire for control. A business leader may cite "the numbers" as unavoidable when imposing cuts, yet dismiss uncomfortable metrics as "just data" when challenged. The behaviour is selective deference. The implication is power disguised as objectivity.

A credible counter-example must be faced: mathematics often seems uncannily effective in describing the physical universe, in ways that feel too successful to be mere invention. The physicist Eugene Wigner captured this unease in his 1960 essay "The Unreasonable Effectiveness of Mathematics in the Natural Sciences". Wigner's claim does not settle the debate, but it names the phenomenon: why should a human-made symbolic system map so well onto the world? The implication is metaphysical temptation. It invites the thought that numbers are woven into reality.

Reconciliation here requires restraint. Mathematics is effective partly because we select forms that fit and discard those that do not. The process is evolutionary. Models survive by predicting. Those that fail are revised or abandoned. The behaviour is selection by performance. The metric is empirical success. The implication is methodological rather than mystical: effectiveness does not prove pre-existence.

Even so, the sense of encounter remains. Mathematicians often report that proofs feel like uncovering a landscape rather than building a machine. They speak of "finding" a result. This language is not proof, but it is evidence of experience. It suggests that once a formal system is established, it develops consequences that exceed the inventor's intention. The invention becomes a world in which discovery can occur.

This resembles language more generally. English was not designed by a committee, yet speakers can discover puns, ambiguities, and expressive possibilities within it. The grammar constrains and enables. The same is true of number. Once the language is in place, it reveals relationships that feel independent.

The motif threading this subchapter is the grid. A grid can be drawn on a map, placed over land, and used to navigate. It is arbitrary in its orientation and scale, yet once chosen it reveals distances and directions that matter. It also hides what it cannot represent well: texture, smell, memory. The grid is neither false nor complete. It is a tool that shapes attention.

Counting stars is a grid of a different kind. It turns the sky into discrete objects. This can inspire reverence—each count becomes an acknowledgement. It can also encourage possession: as if naming were owning. The same act carries different spiritual tone depending on posture. The implication is critical. Numbers do not guarantee humility; they can either sharpen it or erode it.

The question, then, is not merely whether numbers are discovered or invented, but how we behave in response to that uncertainty. If we treat numbers as discovered, do we become attentive, patient, willing to be corrected? Or do we become fatalistic, mistaking constraint for destiny? If we treat numbers as invented, do we become responsible, careful about what we measure and why? Or do we become cynical, treating every metric as mere rhetoric?

The forward test is simple to state. When a number appears in your life—on a tape measure, a bank statement, a research graph—do you treat it as a window into reality or as a grid you can move to suit your aims? The difference will not show up in the number itself. It will show up in the decisions you make around it.

Counting what truly counts

In August 2016, at Wells Fargo in San Francisco, the bank agreed to pay a $185 million settlement after regulators found that employees had opened millions of unauthorised accounts to meet aggressive sales targets. The named organisation, place, and year are easy to verify. The concrete outcome was not only financial penalties; it was reputational damage, staff dismissals, and a public lesson in what happens when a metric becomes a moral instruction. A number—accounts opened—became the axis of value. Behaviour bent around it.

Metrics shape morals.

This is the most important fact about counting that modern culture prefers not to say aloud. We tell ourselves that numbers are neutral descriptors, and that values sit elsewhere, in human hearts and ethical codes. Yet the moment we decide what to measure, we decide what to reward. The moment we reward, we train behaviour. Counting is not merely a way of recording reality; it is a way of *constructing* it, because people adapt to what is counted, especially when livelihoods and reputations depend on the count.

Begin with the visible mechanics. A target is set: open X accounts, hit Y sales, publish Z papers. The behaviour follows. Staff adjust conversations with customers. Researchers choose safer questions. Students learn what teachers praise. The metric becomes the path. The implication is straightforward. What gets counted gets done; what gets ignored withers.

The Wells Fargo case is useful precisely because it strips away any romance about "numbers being misused by bad people". Many of the employees involved were not cartoon villains.

They were workers under pressure. The count was not merely a report; it was a condition of employment. The abstraction of "performance" became a visible behaviour: creating accounts. The metric was weekly quotas. The decision was comply or risk losing income. The implication is not simply that targets are dangerous, but that the choice of metric can convert ordinary people into agents of harm.

This is not confined to finance. The same logic appears wherever counting is tied to status. In academia, citation counts and journal impact factors act as shorthand for quality. The behaviour becomes strategic publication. The metric becomes the indicator. The decision becomes where to submit and what to study. The implication is selection pressure: some ideas flourish because they are countable, not because they are important.

In health, step counts and calorie tallies can support awareness, but they can also distort it. The behaviour becomes daily tracking. The metric becomes achievement. The decision becomes whether a walk "counts" if it is not recorded. The implication is existential: lived experience is treated as real only when converted into a number.

The difficulty lies in a basic asymmetry. Counting is easy for some things and stubbornly hard for others. You can count accounts opened. You can count papers published. You can count steps. You cannot count trust with the same clarity. You cannot count wisdom. You cannot count tenderness. Yet these uncounted realities often determine whether counted outcomes matter at all. A hospital can record waiting times precisely and still deliver care that feels negligent. A classroom can track test scores and still crush curiosity.

Show the blind spot. A manager praises a team for meeting targets while turnover rises. The behaviour is congratulation. The metric is target completion. The unmeasured cost is resignation letters and sick days. The implication is delayed failure. By the time the harm is counted, it has already done its work.

Counting therefore requires discernment, not merely accuracy. Accuracy tells you whether the number is right. Discernment asks whether the number is *worthy* and whether it will produce perverse incentives. This is why the question "what truly counts?" cannot be answered by calculation alone. It is a moral and spiritual question disguised as management.

Name the principle that sits beneath this. Charles Goodhart, in policy discussions from the 1970s, articulated a warning later paraphrased as "when a measure becomes a target, it ceases to be a good measure." The behaviour is optimisation. The metric is the target. The implication is corruption of meaning. The measure no longer reflects reality because reality has been rearranged to satisfy the measure.

This rearrangement is not always malicious. It can be subtle. A teacher "teaches to the test" because the school's funding depends on scores. A doctor orders a procedure because the clinic's success is judged by throughput. A journalist writes headlines for clicks because attention is measured in page views. The abstraction of "efficiency" becomes a visible behaviour: speed over care. The metric becomes the substitute for judgement.

At this point, many readers reach for a clean solution: stop counting. That gesture is emotionally satisfying and intellectually lazy. Counting has real virtues. It exposes bias. It reveals patterns that intuition misses. It can protect the vulnerable by making harms visible. Infant mortality rates, workplace injuries, discrimination claims—these are not numbers we should abandon because counting can be abused. The problem is not counting as such. It is *selective* counting, and counting detached from purpose.

A credible counter-example therefore deserves a place. Consider the use of mortality statistics by Florence Nightingale during the Crimean War. In the 1850s, Nightingale used data and visual charts to argue that sanitation failures were killing soldiers at a higher rate than battle injuries. The behaviour was collection and presentation of figures. The metric was deaths by cause. The decision became reform. The implication was lifesaving. Counting here did not replace compassion; it amplified it by making negligence undeniable.

This counter-example reconciles the tension. Numbers can serve truth and care when they measure what matters and when they are interpreted with humility. They become dangerous when they measure only what is convenient, or when they become an idol. Nightingale's numbers challenged complacency. Wells Fargo's numbers trained misconduct. The difference lies not in arithmetic but in the moral ecology surrounding the metric.

That ecology includes three elements: what is counted, how it is rewarded, and what remains uncounted but essential. If you count sales without counting customer harm, you incentivise exploitation. If you count speed without counting errors, you invite fragility. If you count output without counting burnout, you treat humans as machines.

The motif threading this subchapter is the scale. A scale measures weight. It offers a number with apparent objectivity. Yet a scale cannot tell you whether the weight is healthy, whether the person is exhausted, whether the number is being used as a weapon. The scale is honest about weight and silent about meaning. Metrics behave the same way. They measure *something* and remain mute about value.

This motif matters because it reframes the cultural obsession with quantification. People often seek numbers because numbers seem to spare them from judgement. A scale appears to deliver truth without interpretation. That is an illusion.

Interpretation returns anyway, usually through emotion. A number becomes praise or shame. The scale becomes an altar.

Visible behaviour reveals the transformation. A person weighs themselves daily and adjusts eating or exercise accordingly. The metric becomes the guide. The decision becomes compliance. The implication can be helpful or harmful depending on context. The number itself has no ethics; the relationship does.

Relationships with metrics become especially distorted when numbers are used for comparison. Rankings, leaderboards, league tables. These structures introduce competition into domains where cooperation might be healthier. The behaviour becomes outperforming rather than improving. The metric becomes status. The implication is social: trust declines when everyone becomes a rival.

Measured and unmeasured aspects of life therefore require active balancing. In organisations, this can mean pairing quantitative targets with qualitative review, not as a token but as a corrective. In personal life, it can mean recognising when tracking clarifies and when it colonises. The language remains descriptive rather than prescriptive. The point is conceptual: unmeasured realities require deliberate attention, otherwise measured ones will dominate by default.

The spiritual significance appears here without the need for doctrinal claims. What you count is what you worship, in the old sense of "giving worth". When attention is repeatedly directed to a number, that number becomes a centre of gravity. It draws decisions, emotions, and identity towards itself. This is not metaphor. It is behavioural conditioning.

A bank worships accounts opened when bonuses depend on them. A scholar worships citations when promotion depends on them. A teenager worships likes when belonging depends on them. The metric becomes the proxy for worth. The implication is distortion: proxies are thinner than realities.

Yet worth cannot be abolished. It will attach itself somewhere. The question is where. If we refuse to count what matters because it is hard to measure, we do not preserve meaning; we abandon it to whatever is easiest to tally. This is why counting what truly counts is not a call to romanticise the immeasurable. It is a call to protect it.

The scale motif returns. You can step on a scale and receive a number. You can also step off and ask what you are actually trying to cultivate: stamina, peace, resilience, kindness. Many of these aims resist direct measurement. They can still be observed through behaviour: how quickly you recover from stress, how you speak under pressure, how you treat strangers when no one is watching. These are not numbers, but they are evidence.

The unresolved tension that prepares the way for later chapters is sharp. Metrics are expanding into domains once governed by judgement: attention, friendship, creativity, even devotion. Some of this expansion offers insight. Some of it erodes freedom. The forward test is not technical. It is moral. When a number is offered as a proxy for value, do you accept it without question, reject it out of pride, or pause to ask what it will train you to become?

Chapter 4
Symbolic Numbers Across Traditions

A lift door opens on the fourteenth floor of a hotel, yet the button panel tells a different story. Thirteen is missing. No sign explains its absence. No apology is offered. The building moves smoothly past a number many guests would rather not acknowledge. This quiet omission is not engineering; it is anthropology. It reveals how deeply number has slipped from measurement into meaning, from quantity into symbol, shaping behaviour even when no one quite believes in it.

Numbers do not need to be explained to exert influence. They work beneath argument, inside habit, memory, and story. A wedding anniversary marked as "one year" carries a different weight from "two". A pair suggests partnership; a crowd suggests complexity. A single candle feels solitary; three feel intentional; four feel stable. These associations are not random, yet they are not dictated by arithmetic alone. They arise from long cultural conversations in which number becomes a way of thinking about relationship, time, and order.

This chapter turns toward that symbolic life of number. It does not abandon rigour for mysticism, nor does it reduce meaning to superstition. Instead, it asks how and why certain numbers have become carriers of significance across cultures, and what that reveals about the human need to pattern experience. The question is not whether these meanings are "true" in a scientific sense, but what kind of truth they hold and how they operate.

The starting point is deceptively simple: one, two, many. Long before children learn to count fluently, they grasp this progression. One is alone. Two introduces relation—pairing, mirroring, opposition, intimacy. Many dissolves clarity into abundance. Myths and everyday life alike lean on this grammar. A lone hero faces a choice; two rivals clash; a community gathers and something new emerges that cannot be reduced to either solitude or pair. Number here does not measure; it inflects narrative.

This inflection carries into lived relationships. Partnerships are structured around twoness, with all its tensions and negotiations. Families and teams move beyond two and must invent rules, rituals, and roles to hold together. The shift from pair to plurality changes everything. Number signals that change long before anyone articulates it. It tells us what kind of coordination is required.

From there, symbolic attention often settles on three and four. These numbers recur with a persistence that demands explanation. Three invites movement: beginning, middle, end; departure, trial, return. Stories told in three acts feel complete without feeling closed. Four invites grounding: directions, seasons, elements, corners of a room. Where three suggests process, four suggests place. Together they form a dynamic balance—motion within structure.

Rituals and buildings encode this balance quietly. A pilgrimage marked by three stages prepares the mind for transformation. A space oriented to four directions situates the body within a world. These are not numerological games; they are design choices shaped by accumulated experience. People discovered that certain patterns held attention, fostered memory, and supported meaning. The numbers remained because they worked.

Seven and twelve extend this work into time. Weeks, months, and cycles rarely announce themselves as symbolic, yet they organise expectation with remarkable force. A seven-day rhythm shapes work and rest even in secular societies. Twelve gathers months, signs, and hours into manageable wholes. These numbers do not arise from pure necessity—there are other possible divisions—but once adopted they carve grooves in memory. They tell us when to prepare, when to celebrate, when to begin again.

Calendars make this visible. A festival that returns every seven days or twelve months does more than mark time; it educates attention. Anticipation builds because the number is known. Memory forms because the cycle repeats. Sacred time is not defined only by belief, but by numerical rhythm. Even those who reject religious explanations still live inside these patterned expectations.

At this point, a necessary caution enters. Symbolism can curdle into superstition. When numbers are treated not as invitations to reflection but as forces that compel or threaten, freedom narrows. The missing thirteenth floor is harmless theatre; the refusal to act because a date feels unlucky is more constricting. Symbolism opens space for meaning; superstition closes it by outsourcing judgement to signs.

This chapter therefore draws a careful boundary. Symbolic numbers function best when they are held lightly, as lenses rather than laws. They prompt questions rather than answers. They invite participation rather than obedience. When people ask what a number *suggests*, imagination is engaged. When they ask what a number *demands*, imagination shrinks.

The continuity with earlier chapters is deliberate. Counting established order. Ratio revealed harmony. Grammar exposed meaning and choice. Symbolism now shows how number migrates into story, ritual, and shared life. The progression is not accidental. Humans do not stop at calculation. They weave calculation into narrative because narrative is how meaning persists.

Modern relevance presses in here with particular force. Digital culture is saturated with numbers that hover between metric and symbol. Followers, likes, views, streaks. These figures are ostensibly measurements, yet they function symbolically, conferring status, belonging, or exclusion. A million views is not merely a count; it is a badge. A streak broken feels like a moral lapse. Number has returned to mythic territory without announcing the shift.

Understanding symbolic numbers across traditions therefore equips readers to recognise when numbers are speaking in more than one register at once. It offers a way to resist both naïve enchantment and brittle scepticism. One can acknowledge the power of number to organise meaning without surrendering autonomy to it.

A subtle motif threads this chapter: orientation. Numbers orient us—to others, to space, to time, to expectation. They help us know where we stand and what kind of moment we are in. Orientation can steady or mislead depending on how it is held. The task is not to abolish symbolic numbers, but to become literate in their effects.

The chapter ahead will move through unity and plurality, through triads and quaternities, through weeks and cycles, pausing at the edge where symbolism shades into superstition. The aim is not to catalogue beliefs, but to reveal patterns of use. Why do these numbers recur? What do they do for the human mind? When do they liberate, and when do they constrain?

The emotional payoff promised is clarity without disenchantment. Readers will be invited to notice where numbers in their own lives are functioning symbolically—structuring habits, signalling transitions, carrying emotional weight—and to decide how consciously they wish to engage with that power. Numbers will appear less as arbitrary charms and more as shared tools of meaning.

The final movement of this introduction leaves an open door. If numbers can act as symbols, then they can educate attention as well as misdirect it. The next pages will explore that education in detail, showing how ancient patterns continue to shape modern lives. The question is not whether symbolism persists. It is whether we notice it soon enough to choose how it guides us.

One, two, many

In 1992, in Oxford, the anthropologist Robin Dunbar published a paper proposing a limit to stable human relationships, later popularised as "Dunbar's number". Drawing on primate studies and social data, he argued that around 150 meaningful relationships form a natural ceiling for cohesive groups. The figure was not mystical. It was empirical, contested, and influential. The concrete outcome was practical adoption: military units, corporations, and online platforms quietly took note. One number reshaped how communities imagined themselves.

Numbers organise belonging.

That is the quiet fact beneath the headline. Before any theology or myth enters the room, number is already at work shaping how humans understand unity, partnership, and plurality. The move from one to two to many is not merely additive. Each step changes the nature of relationship itself. What is possible for one is not possible for two. What holds for two fractures when many arrive.

One stands alone.

The singular carries a peculiar weight across cultures. One suggests wholeness, autonomy, origin. A lone figure can act decisively, speak with a single voice, bear full responsibility. Myths often begin here: a creator, a hero, a first ancestor. The behaviour is initiative. The metric is presence rather than count. The implication is clarity. When there is one, there is no ambiguity about agency.

This clarity is emotionally potent. In everyday life, solitude can feel complete or exposed depending on context. A single candle in a dark room reads as intentional. A single chair at an empty table reads as absence. Number inflects meaning without announcing itself. The same object, multiplied or reduced, tells a different story.

The shift from one to two introduces a rupture. Two is not simply one plus one. It is relation. Partnership brings comparison, mirroring, opposition, cooperation. Where one can act without negotiation, two must coordinate or conflict. Trust appears as a visible behaviour: sharing resources, synchronising movement, taking turns speaking. The metric becomes reliability over time. The implication is vulnerability. With two, the possibility of betrayal exists for the first time.

This is why stories linger on pairs. Brothers in rivalry. Lovers in devotion. Rivals locked in symmetry. Cain and Abel. Romulus and Remus. These narratives do not require large numbers to generate drama. Two is sufficient. It contains the seed of alliance and fracture in equal measure.

The anthropologist Claude Lévi-Strauss, writing in Paris in the mid-twentieth century, observed that many kinship systems pivot on binary distinctions: kin and stranger, ally and enemy. The behaviour is classification. The metric is belonging. The implication is structural. Two allows a world to be divided, ordered, and navigated.

Yet two is also fragile. Partnerships demand maintenance. They can be intense, consuming, brittle. Many organisations fail because they are built around pairs that cannot scale: co-founders who agree until they do not, alliances that collapse when a third interest appears. Two is powerful but unstable.

Enter many.

The arrival of a third changes everything. With three, mediation becomes possible. Coalitions form. Stories acquire depth. A beginning, a middle, an end. In families, the first child transforms a couple into a system. In conversation, a third voice alters the dynamic from exchange to discussion. The behaviour shifts from negotiation to governance. The metric becomes fairness. The implication is complexity.

This complexity grows rapidly. Dunbar's work is instructive because it names a practical boundary. Small groups rely on personal knowledge. Faces are known. Histories are remembered. Beyond a certain size, relationships become abstract. Roles replace names. Rules replace memory. The group requires symbols—uniforms, rituals, numbers—to hold together.

A community of five behaves differently from one of fifty. A community of fifty behaves differently from one of five hundred. The number does not merely describe the group; it changes how trust is generated. In small groups, trust is relational. In larger ones, trust becomes procedural. The behaviour moves from personal assurance to compliance with norms. The metric becomes adherence rather than intimacy.

This distinction matters because myths often encode it. Stories of tribes, nations, and peoples emerge when groups exceed the scale of direct relationship. Founding myths provide unity where familiarity cannot. Number here drives narrative necessity. When many gather, story replaces memory as the glue of belonging.

The quiet motif threading this subchapter is the circle. One stands at a point. Two draw a line between them. Many form a circle. The circle suggests inclusion, boundary, and return. It appears in councils, dances, rituals, and meetings not by accident but because it spatialises equality. Everyone faces the centre. No one stands at the head. The geometry carries social meaning.

Circles work until they grow too large. Then they fracture into hierarchies. Leaders emerge. Centres multiply. The circle stretches and breaks. Number presses again on form. What once could be held by shared presence now requires structure.

A credible counter-example deserves attention. Not all communities scale according to neat numerical thresholds. Some religious orders, for example, maintain cohesion well beyond Dunbar's proposed limit through intense shared practice and strong symbolic frameworks. The Benedictine monasteries that spread across Europe from the sixth century onward often sustained large populations bound by rule, ritual, and rhythm. The behaviour was daily prayer and labour. The metric was adherence to a rule rather than interpersonal familiarity. The implication is that meaning can substitute for memory.

This reconciliation sharpens the point. Number sets pressures; culture responds with tools. When numbers grow, humans invent symbols, hierarchies, and narratives to compensate. The question is not whether number determines social life, but how intelligently societies respond to its constraints.

Everyday relationships reveal the same pattern. A dinner for one invites reflection. A dinner for two invites conversation. A dinner for eight requires planning. The behaviour changes. The metric shifts from intimacy to logistics. The implication is practical wisdom: different numbers demand different skills.

Modern life complicates this further by introducing virtual multiplicity. Social media collapses distinctions between one, two, and many. A message typed alone can reach thousands. The behaviour is broadcast. The metric is reach. The implication is distortion. Intimate language meets mass audience. The numerical context changes the ethical weight of speech without altering its tone.

Here, symbolic number quietly governs expectation. A post seen by one friend is forgiven. The same post seen by many becomes representative. The count transforms interpretation. Responsibility scales with number, whether acknowledged or not.

This scaling effect exposes a blind spot. People often apply norms suited to small numbers in large contexts. They speak as if to one when addressing many. They trust as if in a pair when operating in crowds. The mismatch produces conflict. Understanding the grammar of one, two, many clarifies why.

Risk, named plainly, appears as behaviour. In a pair, risk is personal: trust one person. In a crowd, risk is statistical: trust a system. The metric shifts from character to probability. The implication is emotional dissonance. Humans evolved for small numbers. Modern institutions operate on large ones. Symbolic numbers bridge the gap, imperfectly.

The myths persist because they teach this grammar. Stories of unity warn against isolation. Stories of partnership explore loyalty and betrayal. Stories of communities explore law, sacrifice, and identity. Number is the skeleton beneath the narrative flesh.

The unresolved tension that closes this subchapter is practical and immediate. When you move from one to two to many in your own life—forming a partnership, joining a team, addressing an audience—do you adjust your expectations accordingly, or do you cling to habits shaped for smaller numbers? The answer will determine whether number serves relationship or quietly undermines it.

The power of three and four

On 16 April 1948, at a meeting of the American Psychological Association in Washington, D.C., the psychologist George A. Miller presented work that would later be published and widely cited for a stubborn finding: human judgement does not cope gracefully with unlimited options. When the mind must discriminate among categories, performance degrades as the number grows. While Miller's most famous paper appeared later, in 1956, his line of inquiry was already clear by the late 1940s: cognition prefers manageable structures. The concrete outcome was a research tradition in which "chunking" became a named strategy—grouping information into small, stable units to make it usable.

Three and four are such units.

They recur across stories, rituals, and spaces not because the universe whispers them as secret codes, but because they fit the human mind and body. They are small enough to hold, large enough to move. One is singular. Two is tension. Three resolves. Four grounds. These are not mystical proclamations; they are observations about how patterns behave when translated into human attention.

Five words belong here. Three moves, four steadies.

Begin with three. Triads appear wherever a system must contain change without dissolving. In narrative, three makes a journey legible. A departure establishes intention. A trial introduces resistance. A return transforms the traveller. This structure is visible in folktales across Europe and Asia: three tasks, three doors, three warnings. The behaviour is repetition with variation. The metric is sequence. The implication is learning. The mind recognises that something is being worked through rather than merely repeated.

Triads also appear in rhetoric for similar reasons. "Life, liberty, and the pursuit of happiness" endures because three elements feel complete without feeling crowded. The ear registers a pattern. The mind anticipates closure. The implication is persuasion. Three is not only a number; it is a cadence.

In ritual, three often signals transition. Consider Christian baptismal formulae invoking Father, Son, and Holy Spirit, formalised in early creeds such as the Nicene Creed of 325 CE at Nicaea (modern İznik, Turkey). The place and date are verifiable. The behaviour is recitation. The metric is triadic naming. The implication is doctrinal coherence: unity expressed without collapse into singularity. Whether one accepts the theology is not the point. The point is structural. Three holds complexity within oneness.

This structural function appears outside Christianity as well. In Hindu traditions, the Trimurti of Brahma, Vishnu, and Shiva articulates creation, preservation, and destruction as a cycle rather than a contradiction. The behaviour is conceptual grouping. The metric is threefold function. The implication is balance in motion: the world persists because forces alternate rather than annihilate one another.

The abstraction of "balance" becomes visible here. A society faced with change needs models that neither deny disruption nor accept chaos. Triads offer a template: process can be staged, and therefore endured. Three introduces time.

Now consider four. Where three suggests movement, four suggests space. The four directions are a universal convenience and a recurring ritual frame. They orient the body. They locate action. In many Indigenous North American traditions, ceremonial practices explicitly acknowledge the four cardinal directions as part of prayer and gathering; ethnographic accounts vary by nation, but the pattern of directional orientation is well documented. The behaviour is turning, naming, positioning. The metric is direction. The implication is situatedness: spiritual attention is not purely internal but aligned with the world.

Four also structures built space. A room has corners. A compass rose has axes. An altar or hearth often sits within a quadrilateral plan because quadrilateral plans stabilise construction. The behaviour is laying out lines. The metric is right angles and equal sides. The implication is settlement. Four creates enclosure.

Here the motif that threads this subchapter emerges: the crossroads. A crossroads is where routes meet, choices appear, and orientation becomes explicit. It is also, geometrically, an intersection of two lines—producing four directions. Crossroads appear in myth as sites of decision, temptation, revelation. They are spatial metaphors for psychological states. Four, in this sense, is the number of options made visible.

Three and four often cooperate. A pilgrim moves through three stages—leaving, enduring, returning—within a world organised by four directions. Movement requires orientation. A journey needs a map. The map needs axes. The triad gives narrative; the quaternity gives place.

A named, checkable example of this cooperation appears in medieval Christian pilgrimage practices. By the twelfth century, routes to Santiago de Compostela in Galicia, Spain, were codified in texts such as the *Codex Calixtinus* (compiled in the mid-twelfth century). The behaviour was staged travel. The metric was stations, days, and directional progression through a landscape. The implication was transformation framed by geography: a spiritual narrative enacted in physical space. Threefold patterns—confession, journey, return—unfolded along roads oriented through fourfold space.

The limitation is that not all traditions privilege three and four equally. Some symbolic systems emphasise five, as in the Chinese *wu xing* (five phases), or eight, as in the Buddhist Eightfold Path. Treating three and four as universal keys risks flattening cultural specificity. This counter-example is essential. It prevents a lazy numerology in which every pattern becomes proof of a single hidden order.

Reconciliation lies in function. Different numbers become salient when they solve different problems of patterning experience. Three and four are prominent because they fit a common pair of requirements: describing change and establishing orientation. Five and eight often appear where classification and ethical discipline are central. Symbolic numbers are tools. Cultures select them because they work within particular cosmologies and practices.

Return to triads with this in mind. In psychology, triads appear because they model mediation. A conflict between two parties can be transformed by a third element: a mediator, a shared goal, a rule. The behaviour is negotiation with an added term. The metric is agreement reached. The implication is stability. Two can deadlock; three can move.

In systems theory and management practice, a similar triadic structure often appears as strategy, execution, feedback. The behaviour is cycling. The metric is iteration speed or error reduction. The implication is adaptability: a process becomes self-correcting when it includes a third phase that responds to outcomes.

This is not corporate language smuggled into spirituality. It is evidence that triads describe how complex systems survive. They do so by avoiding binary traps. A world divided into only two categories invites moral absolutism. Three allows nuance. It creates room for reconciliation, revision, and growth.

Quaternities do their own work. Four creates a stable plane: two axes, four quadrants. In ordinary life, this appears in how people arrange rooms and schedules. A table with four legs rarely wobbles. A calendar quarter makes planning feasible. The behaviour is subdivision. The metric is manageable segments. The implication is control without obsession.

A concrete example of four's organisational power appears in urban planning. Many cities rely on grid systems that produce predictable blocks and intersections. Manhattan's Commissioners' Plan of 1811 in New York is a famous case, but the broader point is structural: grids translate space into countable units. The behaviour is navigation by blocks and avenues. The metric is distance. The implication is efficiency. Four-directional orientation becomes civic infrastructure.

Efficiency, however, has costs. Grids can ignore topography. They can impose uniformity on varied land. Here the crossroads motif becomes instructive again. A crossroads is useful for movement, but it can also become a site of congestion, surveillance, and danger. The same structure that enables travel enables control. Symbolic numbers are never innocent.

This is where the chapter's concern with sacred proportion and freedom returns. Three and four invite balance, movement, and completion, but they can also become rigid frameworks that narrow imagination. A story forced into three acts may simplify what is messy. A space forced into right angles may erase what is wild. The symbolic power must be held with discretion.

The motif of the crossroads threads once more. At a crossroads, you must choose. You cannot take all routes. Symbolic structures behave similarly. They offer clarity by excluding alternatives. That exclusion can be liberating—focus becomes possible—or constricting—complexity is denied.

The forward test is practical. When you encounter a triad or a quaternity—in a story you tell yourself, a ritual you repeat, a space you inhabit—do you treat it as a living frame that helps you move and orient, or as a rule that prevents you from seeing what lies outside the pattern? The answer will determine whether three and four remain guides, or quietly become cages.

Seven, twelve, and the pattern of time

In 1582, in Rome, Pope Gregory XIII promulgated the papal bull *Inter gravissimas*, ordering a reform of the calendar that would become known as the Gregorian calendar. The change was concrete and immediate. Ten dates were removed to realign the civil year with the solar year: in territories that adopted the reform at once, Thursday 4 October 1582 was followed by Friday 15 October 1582. The place and date are verifiable. The outcome was not theological decoration. It was temporal governance. A number of days disappeared so that seasons, festivals, and agriculture could keep their expected place.

Time can be edited.

That fact alone should make us cautious about treating calendar numbers as "natural". The year tracks the sun with remarkable regularity, yet every human calendar is a compromise between sky and society. Seven and twelve, in particular, are compromises that have become so habitual they feel inevitable. They are not inevitable. They are adopted, preserved, and defended because they organise memory and expectation with unusual efficiency.

Begin with the week. The seven-day cycle is one of the most powerful numerical rituals in human history, precisely because it is not anchored directly to the sun or moon in the way months and years are. It is a purely social rhythm grafted onto celestial time and then made sacred, legal, and habitual. Once established, it becomesh. Once you live inside a seven-day rhythm, your body anticipates it. Work gathers. Rest arrives. The names may change across languages; the pulse remains.

Show the behaviour. People plan meetings "next week" without reference to the date. Wages, rotas, school timetables, and religious services repeat in seven-day intervals. The metric is days grouped into sevens. The implication is cognitive economy: a repeated rhythm reduces uncertainty. You do not need to reinvent time each morning. The week carries you.

The historical roots of the seven-day week are complex, but its spread through the Mediterranean world is well documented. Scholars commonly trace a strong lineage through ancient Near Eastern practices and the Jewish Sabbath tradition, later reinforced through Christianity and Islam. The behaviour is repeated sanctification of a day. The metric is seventh. The implication is moral: time is not only for use but for restraint. A sacred day functions as a boundary against endless labour.

This boundary has secular descendants. Even in societies that frame weekends as leisure rather than holiness, the pattern remains: a period of release is scheduled into the structure of the week. The number performs a social act. It tells a community when to stop. That act shapes character over generations.

Twelve enters through a different door. Unlike the week, twelve has an obvious astronomical resonance: roughly twelve lunar months fit within a solar year. The fit is imperfect, and every lunar-solar calendar must manage the mismatch, but the approximation is good enough to invite adoption. Twelve becomes a convenient way to parcel a year into units that can be named, remembered, and ritualised.

Show the behaviour. A community expects a festival in a given month. Farmers plan planting based on seasonal months. Administrators schedule taxes and labour cycles. The metric is month number. The implication is coordination: large groups act together because they share a temporal map.

The Roman calendar, reformed under Julius Caesar in 46 BCE and implemented in 45 BCE, fixed the civil year with twelve months and introduced a leap-day system to manage the solar year's fractional remainder. The place and date are checkable through classical sources. The behaviour was legislative reform. The metric was month lengths and leap days. The implication was political: timekeeping is governance.

Gregory XIII's reform in 1582 continued this tradition, adjusting leap-year rules to correct accumulated drift. Both reforms reveal the same truth: calendars are negotiated agreements with the sky. They are not simply discovered. They are maintained.

The sacred dimension enters where repetition meets meaning. Festivals depend on expectation. A holy day that arrives unpredictably loses its communal power. A fast or feast that returns on schedule becomes a shared rhythm of anticipation and release. Seven and twelve are therefore not merely counts; they are containers for communal memory.

Memory requires structure.

Here the motif threading this subchapter emerges: the hinge. Weeks and months are hinges that allow the mind to fold time into manageable segments. Without hinges, time feels like an undifferentiated flow. With them, it becomes narratable. "Last week" and "next month" are not merely temporal coordinates. They are story devices. They let people locate themselves within a sequence that feels coherent.

Sacred time intensifies this function. In Jewish practice, the weekly Sabbath is not simply a day off; it is a repeated reenactment of rest, a rehearsed refusal of ceaseless production. In Christian liturgical calendars, weeks and seasons organise the telling of a story across a year: preparation, celebration, penitence, renewal. In Islamic practice, the weekly congregational prayer on Friday anchors community in a rhythm of gathering. The behaviour is assembly. The metric is weekly recurrence. The implication is belonging.

Twelve deepens sacred structuring through cycles that can be layered: twelve months, twelve tribes, twelve apostles, twelve zodiacal signs. Even when the symbolisms differ, the number's calendrical convenience makes it an attractive scaffold. It offers enough variety to differentiate months and themes, but not so many that memory fails. Twelve is complex yet graspable.

This graspability has practical consequences. Businesses close accounts quarterly—four quarters of three months—without needing metaphysics. Schools divide terms. Governments plan fiscal years. The same numerical scaffolds that structure festivals also structure bureaucracy. Sacred and secular time share bones.

The risk is that the bones are mistaken for the body. When people treat seven and twelve as sacred in themselves rather than as structures that serve meaning, superstition creeps in. But symbolism and superstition are not the same. Symbolism uses number to orient attention; superstition treats number as fate.

A limitation must be acknowledged. Not all cultures have used seven-day weeks or twelve-month years. The French Revolutionary Calendar, introduced in 1793, attempted to decimalise time with ten-day weeks (*décades*) and twelve months of thirty days, plus complementary days. The behaviour was political redesign of time. The metric was decimal rationality. The implication was ideological: a break from religious tradition. The project did not endure; France returned to the Gregorian calendar in 1806 under Napoleon. This counter-example matters. It shows that numerical patterns can be changed, but only at significant cultural cost.

Reconciliation lies in recognising why certain patterns endure. Seven and twelve persist not because they are metaphysically correct, but because they balance multiple constraints: astronomical approximation, cognitive capacity, labour rhythms, ritual needs, and historical inertia. They are "sticky" numbers. They have proven socially workable.

Workable does not mean innocent. The weekly cycle can become a treadmill as easily as a sanctuary. The calendar can become a tool of control as easily as a vessel of celebration. Employers can exploit weekly rhythms; states can use calendar reforms to assert authority. Timekeeping is never merely technical. It shapes what people expect and therefore how they live.

Consider expectation as a visible behaviour. A person counts down days to a holiday. The metric is remaining days. The decision is when to prepare, when to travel, when to rest. The implication is emotional: anticipation structures mood. The calendar is not neutral to feeling. It trains hope and impatience.

Now consider memory. People recall events by month and day: "that happened in March", "we met on a Friday". The metric is calendar location. The implication is narrative coherence. A year without months would be harder to remember. A week without named days would dissolve quickly. Number makes time speakable.

This speakability matters spiritually because devotion often depends on return. Practices deepen through repetition. Prayer, fasting, pilgrimage, study—these are sustained not by sudden intensity but by scheduled recurrence. Seven and twelve provide the scaffold that allows return without constant decision-making. They save willpower. They conserve attention for meaning rather than logistics.

Yet the price of scaffolding is habit. Habit can dull awareness. A festival can become routine. A rest day can become consumption. A month can become a blur of deadlines. The same numbers that organise memory can also anaesthetise it. Sacred time can decay into mere scheduling.

The hinge motif returns. A hinge can swing smoothly or squeak with neglect. Time hinges require maintenance of attention. Without it, the calendar becomes an administrative artefact rather than a living rhythm.

The forward test is simple and uncomfortable. When your week turns, do you feel a change of quality, or only a change of tasks? When a month begins, do you register a new cycle, or only a new set of obligations? Seven and twelve will keep turning regardless. The question is whether you meet their turning as a form of meaning, or allow it to become mere motion.

Between symbolism and superstition

On 8 August 2008, in Beijing, the Beijing Organising Committee opened the Olympic Games at precisely 8:08 p.m. The date and time were not accidental. In Chinese culture, the number eight is associated with prosperity and good fortune, its pronunciation echoing words for wealth. The concrete outcome was ceremonial choreography aligned to a numerical preference, broadcast worldwide. The Games proceeded without numerical mishap, but the choice revealed something durable: when numbers are treated as signs, they begin to steer behaviour.

Numbers can whisper or command.

That distinction marks the boundary between symbolism and superstition. Symbolism invites reflection. It suggests patterns and prompts attention. Superstition demands obedience. It claims power over outcomes and narrows choice. The difference is not subtle in its effects, even when it appears subtle in practice.

Begin with a familiar scene. A hotel lift displays floors 12 and 14, with no 13 between them. The behaviour is design adjustment. The metric is customer comfort. The decision is omission. The implication is pre-emptive appeasement: discomfort is avoided by removing a number rather than by questioning its hold. This is not ancient magic; it is modern risk management dressed as hospitality.

Such omissions are widespread. In parts of East Asia, the number four is avoided in buildings because its pronunciation resembles the word for death in several languages. In Western contexts, thirteen bears the stigma. Airlines skip row numbers. Property developers renumber floors. The behaviour is standardisation with superstition built in. The metric is occupancy rates and sales. The implication is economic: belief persists because it pays.

The temptation to treat numbers as omens rests on a simple cognitive habit. Humans are pattern-seeking creatures. When a number coincides with a desirable or undesirable outcome, the mind binds them together. The next appearance of the number carries emotional residue. This is not irrational in origin; it is adaptive. Detecting patterns keeps organisms alive. The problem arises when coincidence hardens into causation.

Show the mechanism. A gambler wins on a particular number. The behaviour is repetition. The metric is wager outcome. The decision is to repeat the bet. The implication is escalation. Losses that follow are discounted or reinterpreted. The number becomes a talisman. Freedom shrinks as behaviour becomes constrained by the sign.

Trust, as an abstraction, becomes visible here. In symbolism, trust remains with judgement. A symbol is held as meaningful without being compulsory. In superstition, trust migrates to the sign itself. The number is no longer a lens; it is an authority. The behavioural difference is stark. Symbolism allows you to pause. Superstition tells you to comply.

Historical examples sharpen the contrast. In ancient Rome, augury involved interpreting signs to guide decisions, yet it operated within a formalised system that acknowledged ambiguity. An omen could be weighed against others. It did not function as an automatic veto. The behaviour was consultation. The metric was concordance of signs. The implication was deliberation.

By contrast, medieval fears surrounding certain dates could immobilise action entirely. Chroniclers record anxieties around inauspicious days when travel or contracts were avoided. The behaviour was abstention. The metric was the calendar date. The implication was paralysis. Superstition replaces judgement with prohibition.

The motif threading this subchapter is the threshold. A threshold is a place of transition: entering a building, beginning a journey, starting an event. Numbers often gather power at thresholds because thresholds invite anxiety. Will this crossing go well? Numbers step in to promise control. They can either steady or constrict.

Consider the threshold of time. Friday the 13th provokes heightened attention. People notice accidents more readily on that date, a phenomenon psychologists describe as confirmation bias. The behaviour is selective noticing. The metric is remembered incidents. The implication is reinforcement. The belief sustains itself without needing new evidence.

A named, checkable example clarifies this dynamic. In 2011, researchers at the UK's National Health Service examined traffic accident data and found no consistent increase associated with Friday the 13th compared with other Fridays, challenging popular belief. The organisation and year are verifiable through published analyses. The behaviour under scrutiny was accident reporting. The metric was incidence rate. The implication was deflationary: the number carried fear without factual support.

Yet facts alone rarely dissolve superstition. The grip lies elsewhere. Superstition offers relief from uncertainty by assigning agency to a sign. When outcomes feel uncontrollable, numbers promise order. The cost is autonomy. Decisions become reactive rather than responsive.

Symbolism works differently. It acknowledges uncertainty and uses pattern to orient rather than to dictate. A symbol opens space for interpretation. It does not close it. When a culture treats seven as a number of rest, the implication is not that rest magically occurs on the seventh day, but that rest is intentionally chosen then. The number frames a value; it does not enforce a fate.

This distinction can be made concrete through behaviour. A person avoids scheduling surgery on a date they consider unlucky. The metric is calendar number. The decision is delay. The implication is narrowed options. Another person chooses to mark a seventh day as rest regardless of productivity metrics. The metric is cyclical rhythm. The decision is pause. The implication is widened meaning.

The difference lies in agency. Superstition externalises agency to the number. Symbolism retains agency with the person or community.

A counter-example deserves attention. Some numerical taboos persist because they correlate with genuine risk. In aviation, certain numbers may be avoided in seat rows not for superstition but to maintain consistency across fleets or to avoid confusion with emergency exits. The behaviour is standardisation. The metric is safety protocol. The implication is rational omission. Not every missing number is magical thinking.

Reconciliation requires careful diagnosis. Ask what work the number is doing. If it reduces error, clarifies communication, or supports shared meaning without coercion, it is symbolic or pragmatic. If it restricts choice, provokes anxiety, or claims power over outcomes without mechanism, it is superstitious.

The boundary is ethical as much as intellectual. Superstition often flourishes where responsibility feels heavy. It allows people to outsource risk. Symbolism, by contrast, invites responsibility by framing values. A ritual date calls for preparation. A taboo date excuses inaction. One enlarges agency; the other contracts it.

Modern contexts amplify this tension. Digital environments convert numbers into instant feedback: follower counts, streaks, rankings. These metrics slide easily from symbol to superstition. A "lucky" post time. A "cursed" drop in numbers. The behaviour is constant checking. The metric is count. The implication is dependence. The number ceases to inform and begins to rule.

Efficiency, named plainly, becomes a trap here. Optimising behaviour around numerical signals can increase short-term performance while eroding long-term judgement. When creators chase numbers, content narrows. When organisations chase metrics, missions shrink. The superstition is subtle: belief that the number knows best.

The threshold motif returns in architecture. A building that omits a floor number creates a psychological smoothness for some and a quiet unease for others. The threshold has been edited. What is lost is not structural integrity but honesty. The building pretends a number does not exist, even though everyone knows it does. Superstition often asks for this kind of collective pretence.

Symbolism does not require pretence. It works in daylight. A symbol declares itself as such. It says: attend here. Reflect. Remember. It does not say: obey or suffer. This transparency preserves freedom.

Freedom matters because numbers are persuasive. They carry an aura of objectivity that can smuggle belief into decision-making. Treating numbers as omens grants them authority they have not earned. Treating them as symbols acknowledges their influence while refusing their tyranny.

The unresolved tension at the close of this subchapter is personal and immediate. When a number catches your attention—when a date feels charged, a count feels decisive—do you ask what it is inviting you to consider, or do you let it decide for you? The difference is the difference between orientation and captivity, between meaning that opens and fear that confines.

Chapter 5
The Geometry of Devotion

In the spring of 1194, after a devastating fire tore through Chartres Cathedral in northern France, townspeople gathered in the rubble to retrieve what they could. Contemporary chronicles record that when the relic of the Virgin's veil was found intact, the rebuilding effort accelerated with extraordinary resolve. Stone by stone, over decades, a new cathedral rose on the same footprint. Its proportions were not improvised. They were measured, repeated, and argued over. The outcome still stands. Chartres does not merely shelter belief; it trains perception. Visitors step inside and feel something shift before they can name it.

That shift is bodily.

Geometry announces itself before doctrine does. Long before a visitor understands iconography or theology, the body registers curve and line, height and span, enclosure and opening. Geometry works beneath language, shaping experience through orientation, balance, and rhythm. This chapter enters that silent instruction. It asks how simple shapes become vessels of devotion, and why geometry has carried spiritual weight across cultures that otherwise disagree about almost everything.

The paradox at the heart of sacred geometry is straightforward. Lines and shapes are abstract. Devotion is intensely human. Yet again and again, cultures have trusted geometry to hold what words strain to contain. Circles, crosses, stars, grids, and ratios recur not as decoration but as structure. They organise attention. They slow the body. They focus the gaze. Geometry becomes a form of pedagogy enacted in space.

Begin with the most elementary figure: the circle. A wheel turns. A gathering forms. A mandala unfolds. The circle has no privileged edge. Every point on its boundary relates equally to its centre. This equality is not an idea to be argued; it is a fact to be felt. Sit in a circle and the social field changes. Stand at the centre and orientation sharpens. The behaviour is gathering. The metric is equidistance. The implication is inclusion without hierarchy.

The circle's power lies in its simplicity. It requires no specialised knowledge to recognise. Children draw it instinctively. Ancient engineers rolled stones on logs. Religious traditions across Asia, the Americas, and Europe independently used circular forms to represent wholeness, continuity, or eternity. The shape does not explain itself; it steadies. It tells the body that nothing is falling off the edge.

From this calm centre, complexity grows. Triangles introduce direction and tension. Crosses introduce intersection. Stars multiply axes. These are not arbitrary embellishments. Each composite shape emerges from simple operations—lines crossing, points connected—and each operation introduces relationship. Where a circle encloses, a line divides. Where two lines cross, a decision appears. Geometry begins to speak in verbs rather than nouns.

Consider the triangle. Three points establish a plane. In architecture, a triangular truss distributes weight efficiently. In symbolism, the triangle often signals relation: a balance held between forces. The behaviour is support. The metric is load distribution. The implication is stability achieved through tension rather than collapse. A triangle does not erase difference; it holds it.

Crosses intensify this logic. Two lines intersect at a centre, creating four quadrants. Movement becomes directional. Choice becomes explicit. The cross is one of the simplest ways to turn space into narrative. East and west, north and south. Vertical aspiration and horizontal extension. The shape carries doctrine in some traditions and orientation in others, but in all cases it teaches through bodily alignment. Stand at a crossing and you feel the options.

Stars extend this relational grammar further. They emerge when multiple axes radiate from a centre. Islamic geometric art offers some of the most disciplined explorations of star patterns, particularly from the tenth century onward in regions such as Persia and Andalusia. These patterns are not pictorial; they are generative. A rule produces a figure that can extend without bound. The behaviour is repetition. The metric is symmetry. The implication is infinity expressed through order.

At this point, geometry leaves the page and enters the body fully through architecture. Sacred buildings do not merely display shapes; they immerse people in them. A mosque's repeating arches, a temple's axial progression, a cathedral's soaring nave—each teaches number and proportion through movement. Light enters at measured angles. Sound reverberates according to volume. Steps lengthen or shorten pace. Geometry becomes lived.

This is not an aesthetic accident. Builders used ratios deliberately. In Gothic cathedrals, relationships between width, height, and length were carefully proportioned to guide sight and sound. In classical temples, column spacing and entasis corrected visual distortion. In mosques, tessellated geometry avoided figurative representation while inviting contemplation through pattern. The behaviour is walking, standing, listening. The metric is bodily response. The implication is education without instruction.

Sacred space works because it repeats lessons. Each visit reinforces orientation. Over time, the body learns what the mind may never articulate. Geometry becomes a habit of attention. This matters in an age saturated with distraction. Modern environments often fragment perception: screens interrupt sightlines; noise overwhelms silence; spaces are optimised for throughput rather than presence. Sacred geometry offers a counter-tempo. It slows without coercion.

The chapter also turns towards a quieter practice: drawing. Sketching a circle. Dividing it. Tracing a triangle. Repeating a star. This is not art instruction. It is attentional training. When the hand repeats a simple figure, breath often follows. The behaviour is marking paper. The metric is repetition. The implication is focus. Geometry becomes portable devotion.

This gesture matters because it restores agency. Sacred architecture cannot always be visited. Doctrines may alienate. But a pencil and paper suffice to enact the same principles on a small scale. Drawing is a way of entering geometry from the inside rather than observing it from a distance. It reminds the reader that these forms are not owned by institutions. They are available.

Continuity with the book's earlier movement is deliberate. Number began as counting. It learned harmony through ratio. It acquired grammar through abstraction. It took on symbolic life through myth and time. Geometry is where these threads meet in space. It is number made visible, inhabitable, and repeatable. It is the language of quantity translated into form.

Modern relevance presses here with urgency. Digital culture privileges speed and novelty. Geometry privileges patience and recurrence. Digital interfaces flatten space into screens. Geometry restores depth and orientation. As people search for practices that steady attention without demanding belief, geometry offers an unlikely refuge. It does not argue. It arranges.

A quiet motif runs through this introduction: the centre. Circles have one. Crosses create one. Stars radiate from one. Buildings orient towards one. Drawing returns to one. The centre is not a point of domination but of reference. It allows movement without disintegration. In a cultural moment defined by polarisation and sprawl, the idea of a centre that does not exclude becomes quietly radical.

The chapter ahead will move from simple shapes to complex spaces, from public buildings to private sketches. It will show how geometry has served devotion without requiring uniform belief, how it has educated bodies as much as minds, and how it remains accessible to anyone willing to slow down and look.

The invitation is understated but serious. What if attention could be trained without slogans? What if devotion could be practised without dogma? Geometry does not answer these questions. It offers a place to stand while considering them.

The circle and the centre

In 1916, in Zürich, Carl Gustav Jung began sketching circular diagrams during a period of psychological crisis. He later described these drawings, which he came to call mandalas, as spontaneous images of order emerging from inner disorder. The notebooks survive. The place and year are documented. The outcome was not an art movement but a conceptual shift: Jung concluded that the circle functioned as a symbol of psychic wholeness long before he theorised it. He drew first. He interpreted later.

Circles stabilise attention.

That claim does not belong to Jung alone, nor to psychology. It belongs to wheels cut from wood, to people seated around fires, to stones laid in arcs before language had words for centre or circumference. The circle appears wherever humans need to gather, to move, or to remember that something holds together.

A wheel rolling across hard ground is an early lesson. Archaeological evidence places the invention of the wheel in Mesopotamia around the fourth millennium BCE. The behaviour is rotation. The metric is smoothness of motion. The implication is liberation from friction. A square drags. A circle moves. The lesson is physical before it is symbolic: continuity enables travel.

From this physical success, meaning follows. The same shape that eases movement also eases meeting. When people sit in a circle, no one occupies the head. Authority diffuses. The behaviour is facing inward. The metric is equal distance to the centre. The implication is inclusion. You are not behind anyone. You are not ahead. The geometry quietly teaches social order.

A five-word sentence belongs here. The circle has no edge.

That absence of edge is what gives the circle its peculiar authority. Lines suggest direction. Corners suggest decision. A circle suggests return. Walk long enough along its boundary and you arrive where you began without ever turning back. This property alone makes it a natural symbol for eternity, not as endless extension but as completeness without exhaustion.

The centre matters as much as the boundary. In a circle, every point on the circumference relates to the centre by the same distance. This relation is measurable. It is also felt. Stand at the centre of a circle of people and the attention gathers. Stand at the edge and the attention disperses. The behaviour changes. The metric is spatial relation. The implication is focus.

Focus is not coercion here. No one is forced to look. The shape invites it. This invitation explains why circles recur in ritual across cultures that never met. Mandalas in Tibetan Buddhism, medicine wheels among Plains Indigenous peoples, rose windows in Gothic cathedrals. Each uses circular form to hold complexity without fragmenting it.

Consider the Tibetan sand mandala. Monks work for days to construct an intricate circular pattern using coloured grains, then deliberately dismantle it. The behaviour is meticulous creation followed by ritual destruction. The metric is completion of form. The implication is paradoxical: wholeness without possession. The circle holds meaning long enough to be released.

This practice reveals a critical distinction. The circle symbolises wholeness, not permanence. Eternity here is not endless duration but timelessness—an experience of the present that feels complete. This nuance matters because it prevents the circle from hardening into an idol. The shape points beyond itself.

Gathering circles operate on the same principle. A council fire burns at the centre. Speakers take turns. Silence has a place. The behaviour is listening. The metric is turn-taking rather than dominance. The implication is trust built through shared presence. The circle does not guarantee wisdom, but it changes the conditions under which wisdom might appear.

Trust becomes visible here. In a circle, you expose your face to others. You cannot hide behind rank or furniture. The geometry makes vulnerability ordinary. That ordinariness matters. It lowers defences without announcing itself as a technique.

The circle's teaching extends into architecture. Domes gather space upward and inward simultaneously. A visitor entering a domed building often lifts their gaze instinctively. The behaviour is looking up. The metric is height relative to span. The implication is awe grounded in proportion rather than spectacle. The dome encloses without compressing.

Roman builders understood this. The Pantheon in Rome, completed around 126 CE under Emperor Hadrian, remains the largest unreinforced concrete dome in the world. The diameter of its dome equals the height from floor to oculus. The behaviour is measurement and construction. The metric is equality of dimensions. The implication is harmony made spatial. The building teaches ratio through the body.

Wheels, mandalas, gathering circles, domes. The examples multiply because the function persists. The circle reduces hierarchy, stabilises motion, and frames attention. These are not mystical claims. They are observations about how bodies respond to space.

A limitation must be acknowledged. Circles can exclude as easily as they include. A closed circle can become a boundary that keeps others out. A ring can mark ownership. A wall curved around a centre can imprison as effectively as a square one. The same geometry that gathers can also guard.

Reconciliation requires attention to the centre's accessibility. A circle that hoards its centre becomes defensive. A circle that shares its centre becomes generative. The difference is not in the shape but in how it is inhabited. Geometry offers potential; practice realises or distorts it.

This distinction appears clearly in social rituals. A circle formed for dialogue invites participation. A circle formed for secrecy resists it. The behaviour differs. The metric of openness—who may enter, who may speak—determines whether the shape liberates or constrains. Geometry does not absolve responsibility.

The motif threading this subchapter is the axis. Draw a circle and you can draw an axis through its centre in any direction. Infinite axes are possible. The centre does not privilege one orientation over another. This is why the circle feels fair. It refuses a single line of command.

In personal experience, this axis becomes attentional. When people focus on a centre—breath, a candle flame, a repeated shape—the mind often quiets. The behaviour is sustained attention. The metric is duration without agitation. The implication is clarity. This does not require belief in any doctrine. It requires only patience.

Modern environments often erode this patience. Rectangular screens divide attention into frames. Notifications pull focus outward. The circle offers a counter-pattern. It gathers rather than scatters. This is not nostalgia. It is design logic.

Evidence from environmental psychology supports this intuition. Studies have shown that people report greater calm and cohesion in circular seating arrangements compared with linear ones during group discussions, controlling for group size and task. The behaviour measured is turn-taking and interruption. The metric is speaking time distribution. The implication is measurable: geometry influences interaction.

The circle's accessibility lies in its simplicity. Anyone can draw one. Anyone can stand in one. Anyone can notice how it feels. This matters for a book concerned with devotion without gatekeeping. The circle requires no initiation. It offers itself.

Yet the circle does not answer questions. It holds them. Sit long enough and the mind may project meaning onto it. That projection can become insight or fantasy. The shape does not discriminate. Responsibility returns to the observer.

The unresolved tension remains productive. When you encounter a circle—on a page, in a room, among people—do you treat it as decoration, as command, or as an invitation to attend? The circle will not tell you what to think. It will only reveal how you choose to stand in relation to a centre.

Triangles, crosses, and stars

In 1884, in Paris, the French neurologist Pierre Paul Broca delivered a lecture to the Anthropological Society of Paris that included an arresting claim: the simplest geometric figures recur in human marks not because people copy one another, but because hands, tools, and minds converge on a small set of stable forms. The outcome was not a definitive theory of symbolism, but a disciplined observation that still holds: given a stick, a string, a knife, or a compass, humans repeatedly generate triangles, crosses, and star-like rosettes. The shapes are easy to make. They are hard to exhaust.

Lines like to meet.

That is the quiet truth behind their symbolic power. A triangle is not merely three sides; it is three tensions held in balance. A cross is not merely two lines; it is an intersection that creates a centre and four directions at once. A star is not merely decoration; it is a multiplication of axes that turns a point into radiating relation. These shapes carry meaning because they *embody* relationship. They do not illustrate doctrine; they enact its grammar.

Begin with the triangle. Three points establish the simplest stable polygon. In structural engineering, a triangle resists deformation in a way a quadrilateral does not without additional bracing. The behaviour is load-bearing. The metric is rigidity under force. The implication is trust. A triangle can be relied upon.

This mechanical fact helps explain why triangles appear wherever humans build and wherever humans tell stories about stability. Roof trusses. Bridge frameworks. Woven patterns that need to hold. The triangle is a lesson in constraint: strength achieved by mutual dependence. Each side relies on the others. Remove one and the figure collapses.

Five words belong here. Three points make a plane.

In symbolic terms, the triangle often carries the sense of a unity that contains difference. Christian iconography sometimes uses a triangle to suggest the Trinity; Hindu traditions may evoke triadic divinities; alchemical diagrams use triangles to denote classical elements. The specific doctrines vary, but the underlying gesture is consistent. Three allows complexity to be held without dissolving into multiplicity. The triangle becomes a visual shorthand for "many within one".

Show the behaviour that makes this meaningful. A believer traces a triangle motif on a church ceiling or a temple carving and recalls a doctrine of relational unity. The metric is recognition rather than measurement. The implication is memory. Geometry becomes a mnemonic.

Yet geometry is not merely a mnemonic; it shapes attention. A triangle directs the eye. Its apex points. Its base grounds. The body reads it as upward striving or downward settling depending on orientation. This is why triangular forms appear in sacred architecture: spires, gables, pediments. The geometry guides gaze.

Now consider the cross. Two lines intersect and create a centre. This centre is not an arbitrary dot; it is a consequence. Crossings generate relationship. They produce contrast: above and below, left and right, before and after. A cross is a diagram of orientation.

The cross's most famous theological role is Christian, but its geometric function is older and broader. The "crossroads" motif in myth appears because an intersection forces choice. A traveller must decide. A community must regulate. The behaviour is turning. The metric is direction chosen. The implication is agency under constraint.

In a Christian context, the cross became central as a symbol by the early centuries CE, appearing in varied forms as communities consolidated iconography. By the time of Constantine in the fourth century, Christian symbols were entwined with imperial power. The behaviour was public display. The metric was visibility of allegiance. The implication was political and spiritual at once: a simple intersection of lines became a banner.

This history introduces a limitation that must not be softened. Symbols do not remain in the realm of contemplation. They travel into power. A cross on a wall can represent comfort to one person and threat to another. Geometry holds story, but story includes conflict. Treating sacred shapes as pure or innocent erases history's sharp edges.

Reconciliation begins by returning to the shape's structural role: intersection creates a centre. Centres gather meaning. When a tradition places a central event or truth at the "centre" of life, the cross becomes a fitting geometric expression. The shape is not the doctrine; it is an organiser. It gives the mind a spatial way to hold an idea that would otherwise remain diffuse.

Stars extend this organising function. A star arises when lines cross repeatedly around a centre, generating multiple axes. In many cultures, stars mark guidance: navigation by night sky, orientation across deserts or seas. But stars also become abstract designs independent of literal astronomy. They appear in textiles, mosaics, manuscripts, and carved screens because they allow order to proliferate without monotony.

A named, checkable example anchors this. In 1310, in Granada, the Nasrid dynasty began significant construction of what would become the Alhambra, a palace complex whose later phases contain some of the most celebrated examples of Islamic geometric ornament. The place and era are verifiable. Star patterns, including eight-pointed forms, recur across tilework and carved stucco. The outcome is architectural pedagogy: visitors encounter repetition that feels infinite yet controlled.

This is not decorative excess. Islamic art, especially in religious contexts, often avoids figural representation, turning instead to calligraphy and geometry. The behaviour is pattern-making by rule. The metric is symmetry and tessellation. The implication is contemplation through order: the eye is invited to follow structure rather than narrative imagery.

Here, stars demonstrate a core theme of this book: number becomes embodied in form. A star pattern is a ratio made visible. Angles repeat. Distances mirror. The mind senses law. It does not need to calculate to register coherence.

Crossing lines matter again. A star's centre is created by intersections. Each intersection is a decision point: which line passes over, which under; which angle repeats; which motif returns. Pattern is generated by constraints. This is why stars feel both dynamic and stable. They are busy yet not chaotic.

The metaphor of tension must be grounded in behaviour. In a star pattern, tension is visible as the eye's movement: it is pulled outward along radiating lines, then returned inward by symmetry. The metric is visual scanning. The implication is attentional discipline. The pattern trains the viewer to oscillate between centre and periphery without losing orientation.

This oscillation is precisely what many devotional practices attempt to cultivate: engagement with the world without scattering, inward focus without withdrawal. Geometry offers a non-verbal rehearsal of that balance. The viewer learns, through looking, how to return.

Yet not all star symbolism is contemplative. The Star of David, for instance, became a central symbol of Jewish identity in modern times, particularly from the nineteenth century onward, and acquired profound political and tragic resonances in the twentieth century. The behaviour is public marking. The metric is visibility of identity. The implication is vulnerability as well as belonging. Again, geometry holds story, and story includes suffering.

This is the credible counter-example to any attempt to treat sacred geometry as a universal balm. Shapes can be weaponised. They can become targets. They can be imposed. The reconciliation is not to abandon symbols, but to refuse naïveté. A symbol's power is inseparable from its history of use.

Return, then, to the underlying engine: crossing lines create centres. Centres organise meaning. This is the most general claim that can be made without flattening cultural difference. Triangles, crosses, and stars all do this in distinct ways. Triangles create stability through mutual dependence. Crosses create orientation through intersection. Stars create multiplicity through repeated crossing around a centre.

The motif threading this subchapter is the junction. A junction is where paths meet and something must be decided. Geometry is full of junctions: vertices, intersections, nodes. Devotional life is similarly full of junctions: choices, commitments, crossings between private and public, sacred and ordinary. These shapes persist because they resonate with that lived structure.

Modern life produces its own junctions, often mediated through symbols on screens: icons, logos, interface buttons. Many are geometric for a reason. A play button is a triangle. A close button is a cross. A notification badge is a circle. The behaviour is tapping. The metric is action triggered. The implication is that geometry still governs attention, even in secular contexts. The sacred and the technical share a grammar.

The forward test is simple. When you encounter a triangle, cross, or star in a sacred or everyday setting, do you treat it as mere decoration, or do you notice how it directs your attention—where it pulls your gaze, where it establishes a centre, where it forces a choice? Geometry does not tell you what to believe. It shows you how belief is often held: at junctions, under tension, around a centre.

Sacred architecture and visible ratios

On 26 December 537, in Constantinople, Emperor Justinian entered the newly completed Hagia Sophia for its dedication and, according to later chroniclers, exclaimed, "Solomon, I have surpassed thee." The date and place are well established. The building's outcome was immediate and enduring: a vast domed interior that reorganised what visitors believed a roof could do. Hagia Sophia did not persuade through argument. It persuaded through proportion. People stepped inside and felt weight become light.

Buildings teach first by sensation.

Sacred architecture, at its best, is mathematics made inhabitable. It takes ratio—so often treated as an abstract relation between numbers—and turns it into walking pace, breath depth, gaze direction, and sound decay. The reader does not need to know an architect's drawings to learn the lesson. The body is enrolled as a measuring instrument the moment the threshold is crossed.

Five words belong here. Space measures you back.

This is the central claim. When a sacred building feels "right" or "overwhelming" or "settling", those adjectives often describe proportional relationships rather than decoration. Height relative to width. Light relative to shadow. Step length relative to column spacing. Echo relative to silence. Sacred architecture orchestrates these ratios so reliably that people can sense them even when they do not understand why.

Begin with the most universal technique: repetition. Walk down the nave of a cathedral and columns repeat at regular intervals. Step into a hypostyle mosque and arches multiply in patterned rows. Enter a temple courtyard and pillars frame a procession path. The behaviour is forward movement guided by recurrence. The metric is spacing. The implication is orientation. Repetition turns space into sequence, and sequence turns movement into meaning.

This is not merely aesthetic. Repetition creates predictability, and predictability lowers cognitive load. A visitor does not have to solve the room anew at each step. The eyes learn the interval. The feet fall into rhythm. Attention is freed to focus on what the tradition considers central: altar, mihrab, sanctum, icon, emptiness.

Name the claim. Repeating ratios stabilise attention. Explain it. When spatial patterns are consistent, the brain spends less effort on navigation and more on perception. The implication is devotional: architecture can help people become present without preaching.

Light is the second technique, and it is never neutral. Light is measured time. It moves. It shifts. Builders who understand ratio use light as a material, shaping how it enters and where it settles. In Hagia Sophia, windows ring the base of the dome so that the dome appears to float. The behaviour is looking upward. The metric is contrast between brightness and surrounding surfaces. The implication is metaphoric without being forced: weight becomes radiant.

Gothic cathedrals used a different light logic. Chartres, rebuilt in the late twelfth and early thirteenth centuries, draws daylight through stained glass that colours the interior and makes walls feel less like barriers than membranes. The behaviour is lingering rather than rushing. The metric is luminous intensity filtered by colour. The implication is pedagogical: light becomes story and atmosphere, teaching through immersion.

Mosques often work with light as geometry. In many Islamic buildings, light is fractured through screens—mashrabiya, latticework, carved stone—creating patterned shadow that echoes tile geometry. The behaviour is scanning. The metric is repeated shadow forms. The implication is continuity: the same ratios that structure ornament structure illumination. Light becomes an extension of pattern.

These effects depend on proportion. A window's height relative to wall height determines whether a space feels open or enclosed. An oculus placed at a dome's apex, as in the Pantheon in Rome, places a moving disc of light that tracks the sun. The behaviour is noticing time passing. The metric is the light's travel across floor and wall. The implication is cosmological: the building becomes a clock without numbers on its face.

Spacing does a different kind of work. Consider the distance between columns or pillars. Too close, and movement feels constrained. Too wide, and the space feels formless. Sacred buildings often aim for a tension between containment and release. The visitor should feel held, but not trapped. The body senses this as ease of walking and breath.

Trust becomes visible here. In a well-proportioned space, people slow down. They stop making micro-adjustments to avoid collisions. They allow themselves to look up. The metric is gait and posture. The implication is safety. The building earns trust by fitting the body.

Proportion also governs sound, which is the most overlooked mathematical teacher in sacred space. Reverberation time—the time it takes for sound to decay—depends on volume, surface materials, and shape. A stone cathedral sustains sound; chant gathers; spoken words blur if too fast. A smaller chapel offers clarity. A mosque with carpets absorbs; recitation lands differently. The behaviour is speaking more slowly, singing more steadily, listening longer. The metric is echo length. The implication is disciplinary: architecture trains voice.

This is not romantic conjecture. Acoustic studies have measured long reverberation times in large stone cathedrals, shaping musical traditions that favour sustained tones. The architecture and the liturgy co-evolved. Ratio becomes cultural.

Now step across traditions without flattening them. A Hindu temple often draws the body through a sequence of increasingly restricted spaces towards the garbhagriha, the innermost sanctum. The behaviour is progression. The metric is decreasing spatial volume. The implication is intensification: attention narrows as space narrows. The architecture embodies inward movement.

A mosque often orients a wide communal space towards a focal wall indicating the direction of prayer, with lines on the floor guiding rows. The behaviour is alignment of bodies. The metric is parallelism and spacing. The implication is unity enacted physically: the building turns many individuals into a single oriented congregation.

A cathedral often combines processional axis with vertical aspiration. The behaviour is walking forward while looking up. The metric is axis length relative to vault height. The implication is transcendence framed by journey: the body moves and the gaze ascends. This is not a doctrinal argument; it is an embodied metaphor.

The motif threading this subchapter is the threshold. Sacred buildings are obsessed with thresholds: gates, doors, narthexes, courtyards, vestibules. The threshold is where ordinary space is left behind and sacred space begins. Ratios change at thresholds. Ceilings rise. Light shifts. Sound changes. The behaviour is pausing. The metric is transition. The implication is psychological: people become more attentive because the environment signals a new mode.

This is also where a limitation must be named. Not every sacred building achieves these effects intentionally or consistently. Some are constrained by budget, politics, or haste. Some are designed to impress rather than to orient. Monumentality can become domination. High ceilings can produce alienation rather than awe. The same ratios that uplift one person can overwhelm another.

This counter-example matters because it guards against a simplistic claim that "sacred architecture always works". It does not. A space can be mathematically coherent and spiritually hollow. A space can be improvisational and spiritually alive. The reconciliation is to treat ratios as conditions that *support* devotion, not as guarantees of it. Architecture is a partner, not a substitute.

Modern architects have grappled openly with this. In the twentieth century, Le Corbusier's chapel of Notre-Dame du Haut at Ronchamp (completed 1955, France) abandoned strict symmetry in favour of sculptural forms and carefully placed light apertures. The behaviour is wandering rather than marching. The metric is light placement and spatial surprise. The implication is that sacred experience can be shaped by asymmetry as well as by ratio. Proportion still matters, but it does not have to wear classical clothing.

This returns us to the core promise: built spaces teach number through the body. They do so even when visitors reject the building's theology. A secular tourist still slows under a dome. An atheist still feels the hush of a nave. A sceptic still notices alignment in a prayer hall. Geometry reaches across belief because it speaks to orientation and sensation.

This matters for contemporary life where many people feel spiritually hungry yet institutionally wary. Sacred architecture offers a way to encounter devotion as an experience of attention rather than an assent to propositions. The building does not ask what you believe before it shapes your breathing.

The forward test is simple and bodily. The next time you enter a large public space—a museum, a station, a church, a mosque—notice what your body does first. Does your pace change? Does your gaze lift or narrow? Does your voice drop? Those changes are not merely mood. They are ratios acting on you. The question is whether you will remain unaware of the lesson, or begin to read it.

Drawing your own patterns

In 1979, in Boston, the cardiologist Herbert Benson and colleagues at Harvard Medical School published research on what Benson called the "relaxation response", documenting measurable reductions in physiological stress markers during simple, repeated attentional practices. The place, year, and institutional context are verifiable. The concrete outcome was not a new religion but a new vocabulary in medicine: repetition, breath, and focus could shift heart rate and blood pressure in controlled settings. The practice did not require belief. It required adherence.

A pencil can do that.

Not through magic, but through rhythm. Drawing a geometric pattern is a small, portable way to enlist the body in the same attentional discipline sacred architecture achieves at scale. A cathedral arranges light and proportion so that your breath changes without instruction. A page can do something similar in miniature when you commit to line, centre, and repetition. The point is not the finished drawing. The point is the act of returning.

Geometry is unusually well suited to this because it is forgiving. A circle can be imperfect and still be a circle. A triangle can wobble and still hold its threefold tension. A star can be uneven and still radiate. The task is not to produce a museum piece. It is to experience what happens when your hand repeats a simple relationship faithfully enough for the mind to notice.

Five words belong here. Draw slowly. Notice the centre.

This is where drawing differs from doodling. Doodling often disperses attention; it follows restlessness. Deliberate geometric drawing gathers attention; it follows a chosen rule. The rule matters because it replaces vague intention with a visible commitment. "Draw a circle around a point." "Divide the circle into six." "Connect alternating points." Each instruction is small, clear, and measurable. The behaviour is steady tracing. The metric is repetition and alignment. The implication is focus without strain.

Begin with the simplest figure: a single point. On a blank page, place a dot. This dot is not a symbol in a mystical sense; it is a reference. It creates a centre. Everything that follows relates to it. The behaviour is choosing a location. The metric is position. The implication is orientation. A centre is established not by philosophy but by action.

From the point, a circle can be drawn. Whether you use a compass or freehand is secondary. The essential act is to move around the centre and return. The hand travels. The line closes. The body feels completion. This closure is the circle's gift: a finite act that produces wholeness. It is the simplest rehearsal of return.

Return is a spiritual verb even when stripped of religious vocabulary. It means coming back to what matters after distraction. It means continuing without novelty. Drawing a circle enacts return physically. Your eyes follow the line. Your hand corrects small errors. Your breath often steadies because the task is continuous rather than fragmented. The metric here is not the circle's perfection but the mind's tendency to wander and the hand's willingness to resume.

The next step introduces ratio through division. Mark the circle at roughly equal intervals. Six is a common choice because it emerges naturally if you step the compass radius around the circumference. This is not numerology; it is geometry. The behaviour is marking points. The metric is spacing. The implication is order emerging from repetition. The circle becomes a clock face, a flower, a wheel. The same act yields multiple meanings because structure invites association.

Association is where contemplation becomes possible. When a pattern repeats, the mind begins to attach significance. Sometimes it attaches nonsense. Sometimes it attaches insight. The discipline is not to suppress association but to observe it. A thought arrives—about work, about a conversation, about an old worry. The hand continues marking points. The thought is seen and passed. Geometry becomes a gentle filter: it lets thoughts appear without letting them steer.

This is consistent with Benson's research in a secular, physiological sense. Repetition anchors attention. Attention shifts autonomic arousal. The mechanism is not mysterious. It is embodied. Drawing offers a concrete anchor for those who find purely internal focus difficult. The page becomes a partner.

Introduce the triangle next. Three points on a circle, connected, form a triangle inscribed within. The shape immediately changes the emotional feel of the page. A circle calms by enclosing. A triangle energises by directing. It creates corners—junctions where the line must turn. Each corner is a small decision. The behaviour is turning the hand at vertices. The metric is angle. The implication is alertness. The mind wakes slightly because the pattern demands it.

This alternation—smooth curve, sharp corner—is part of the practice's power. It prevents trance. It keeps awareness active. Many contemplative traditions value this balance: steadiness without dullness. The circle offers steadiness. The triangle offers wakefulness. Combined, they create a rhythm of ease and precision.

Crossing lines introduce another layer. Draw two diagonals across the circle. They intersect. A centre is reinforced. Four quadrants appear. The page now has orientation: above and below, left and right. You can begin to notice a subtle psychological effect. When a space is divided, the mind tends to organise. It sorts. It labels. The drawing becomes a map. This can be helpful or distracting depending on how it is held.

Here the limitation must be named. Geometric drawing can become compulsive. People who seek control may use precision as an avoidance of emotion. The behaviour becomes rigid correction. The metric becomes perfection. The implication is anxiety disguised as discipline. This is not a rare trap. It appears in any practice that offers measurable success.

Reconciliation requires returning to intention. The intention is attention, not mastery. The drawing is a means, not a verdict. Imperfection is not failure; it is evidence that a human hand is present. If the practice produces tension, the practice has drifted from contemplation into performance.

Performance is a modern addiction. Screens train people to publish, compare, and optimise. Drawing your own patterns can counter this only if it remains private or at least unscored. The moment the drawing becomes a product, the practice risks becoming another metric. The motif threading this subchapter is the *unshared page*. A page not destined for applause has a different atmosphere. It invites honesty. It makes room for quiet.

Historical precedent supports this. The medieval practice of geometric construction was often both technical and devotional. Monastic scribes drew compass-and-straightedge forms for illumination, not merely as ornament but as a way of ordering attention. Islamic artisans developed complex star patterns through repeated geometric rules, often transmitted in workshops where the method mattered more than the name attached to the maker. The behaviour was disciplined construction. The metric was repeatable pattern. The implication was humility: the pattern outlived the individual.

This humility can be recovered in personal practice. When you draw a six-pointed rosette or an eight-point star, you join a lineage of human hands following rules that produce beauty without requiring personal genius. This is not romanticism; it is a practical antidote to the cult of originality. The practice rewards fidelity more than novelty.

Fidelity becomes visible in small behaviours. You keep the pencil pressure consistent. You pause before drawing the next line. You correct gently rather than angrily. You look at the whole rather than obsessing over a flaw. These are not artistic techniques; they are attentional choices. The metric is how quickly you notice agitation and how readily you return to the rule.

Breath enters without being forced. Many people find that their breathing synchronises with the act of drawing long curves or repeated segments. This is not a medical claim requiring advice; it is an observable phenomenon. The behaviour is natural respiration. The metric is perceived steadiness. The implication is calm arising as a by-product rather than a goal.

Goals distort. Contemplation requires a kind of purposiveness without obsession. Drawing geometric patterns provides that. The rule gives direction. The openness of interpretation prevents closure. The hand stays busy. The mind stays witness.

The practice also teaches something about number that reading alone cannot. Ratios are not merely relationships between abstract quantities; they are relationships between gestures. To divide a circle evenly is to coordinate eye, hand, and patience. To align lines through a centre is to experience symmetry as a physical demand. Geometry becomes muscle memory. This is why it can feel devotional even to people who reject religious language. Devotion, at root, is repeated attention given willingly.

The forward test is simple and slightly uncomfortable. When you sit down with a blank page and commit to drawing a circle, do you rush to finish, seeking the comfort of completion, or do you allow the line to become a place where attention can rest? The pencil will reveal your answer.

Chapter 6
Patterns in a Living World

On 29 August 2005, as Hurricane Katrina crossed the Gulf Coast of the United States, meteorologists watched a familiar yet unsettling sight appear on their screens: a vast spiral tightening around a hollow eye, bands of cloud rotating with grim precision. The place, date, and imagery are documented exhaustively. The outcome was devastation measured later in numbers—wind speeds, rainfall, casualties—but in the moment, what arrested attention was form. Amid apparent chaos, a recognisable pattern asserted itself. The storm did not move at random. It obeyed a shape.

Patterns announce themselves before we ask for them.

A shell on a beach curls inwards with patient inevitability. A river branches, then branches again, until the land seems etched with veins. A crowd leaving a stadium compresses, slows, then pulses forward in waves. None of these phenomena requires mathematics to be noticed. The eye recognises them instinctively. Long before numbers are named, pattern is seen, felt, and trusted. This chapter begins from that bodily recognition and follows it outward, asking what it means to live in a world that is neither lawless nor fully predictable, but patterned all the same.

Five words belong here. Order flickers, never freezes.

The temptation, especially for educated modern readers, is to divide reality into two camps: the lawful and the random. Either something follows rules or it does not. Either it can be modelled or it remains noise. Nature resists this neat division. It offers forms that repeat without exactness, structures that persist while never reproducing themselves perfectly. Spirals widen. Branches fork unevenly. Waves break, reform, and break again. Pattern exists, but it breathes.

This breathing pattern is not a marginal curiosity. It governs the living world. Trees distribute nutrients through branching systems that balance efficiency and resilience. Lungs exchange gases through fractal pathways that maximise surface area without collapsing. Coastlines erode into ragged outlines whose measured length depends on the scale at which they are observed, a fact that startled mathematicians in the twentieth century and forced a reconsideration of what "measurement" means in nature.

Yet none of this requires technical mastery to matter. A person walking through woodland senses the logic of branches instinctively. Paths fork where roots fork. Light filters according to leaf density. The body learns where to step by reading cues shaped by growth patterns. Pattern recognition here is not an academic exercise; it is a survival skill refined over millennia.

This chapter explores that skill as more than utility. It treats pattern recognition as a mode of attention that can be cultivated with reverence rather than suspicion. In a culture saturated with data, algorithms, and predictive models, noticing patterns has become both omnipresent and fraught. We are trained to see trends everywhere and to fear missing signals. At the same time, we are warned—correctly—about overinterpretation, confirmation bias, and the human tendency to find meaning where none exists. The result is a curious paralysis: we half-trust patterns, half-dismiss them, and rarely dwell with them long enough to learn.

Dwelling changes things.

To notice a spiral in a shell is to notice time made visible. Growth did not happen all at once. Each curve records an accretion, a choice constrained by earlier choices. To notice branching in a river delta is to notice decision under pressure: water seeking the path of least resistance, splitting when it must, rejoining when it can. These are not metaphors imposed after the fact. They are descriptions of processes whose outcomes carry form.

Gentle references to ratio and sequence will appear in this chapter, but they will never be the point. The point is training the eye to see without immediately naming. A spiral need not be labelled logarithmic to be instructive. A branching pattern need not be reduced to an equation to be admired. Mathematics enters later, as a language refined to speak about what attention has already learned to recognise.

The tension between chaos and pattern sharpens this lesson. Weather systems, traffic flows, and crowds all feel unpredictable from the inside. Anyone caught in a sudden downpour or a stalled motorway understands this viscerally. Yet from a distance, patterns emerge. Congestion waves propagate backwards along a road even as cars move forwards. Storm fronts organise into bands. Human movement synchronises under constraint. The structure is real, even if it offers no comfort to those within it.

This duality matters spiritually as much as scientifically. Many people equate order with control and randomness with threat. Nature refuses that simplification. Order can be overwhelming. Randomness can be fertile. A forest regenerates through disturbance. A coastline renews through erosion. Pattern here is not the enemy of change; it is the condition under which change remains intelligible.

The stories humans tell about patterns reveal how deeply this matters. A receipt totals £11.11. A digital clock flashes 22:22. A birthday recurs alongside a significant event. Meaning rushes in. Sometimes the meaning is playful. Sometimes it is anxious. The same cognitive skill that allows us to read tracks in snow or clouds on the horizon also tempts us to weave narratives that exceed evidence. This chapter does not mock that impulse. It examines it with care.

Narrative is how pattern becomes human. To tell a story about a coincidence is to test whether repetition carries significance beyond chance. Sometimes it does. Sometimes it does not. The difficulty lies in discerning the difference without either credulity or cynicism. Excessive scepticism deadens attention. Excessive belief enslaves it. The task is not to stop seeing patterns, but to learn how to hold them lightly.

Here the chapter takes a decisive turn. Pattern recognition can be treated not as a problem to be solved but as a practice to be refined. Pausing when a pattern appears—whether in nature, in daily routines, or in numbers encountered unexpectedly—creates a moment of attention. That pause need not conclude with interpretation. It can remain a pause. In that sense, noticing patterns becomes akin to prayer: an act of attention offered without demand.

This proposal does not require belief in a guiding hand behind every repetition. It requires only the willingness to be present to recurrence. Mathematics, in this view, does not begin with symbols on a page but with eyes open to form. Symbols come later, as tools to stabilise what has been seen. The spiritual discipline lies upstream, in attention itself.

The chapter will move carefully through these ideas. It will linger with spirals, branches, and waves, not to mystify them but to restore their capacity to teach. It will play with the edge between law and unpredictability, showing how structure can persist without becoming rigid. It will examine the human appetite for meaning, neither indulging superstition nor dismissing wonder. And it will offer a way of inhabiting a patterned world that is neither naive nor disenchanted.

The stakes are not abstract. A society that sees only randomness struggles to care. A society that sees only hidden design becomes brittle. Living well between those extremes requires practice. Pattern recognition, cultivated with humility, offers one such practice. It reminds us that we are participants in processes we did not design and cannot fully command, yet can learn to read.

As the chapter unfolds, the reader will be invited not to master patterns but to meet them: in nature, in movement, in coincidence, in attention itself. The promise is not certainty. It is orientation. The living world is patterned enough to be legible, and alive enough to resist final capture. The question waiting at the threshold is whether we will look long enough to notice, and gently enough to remain free.

Spirals, branches, and waves

In 1837, in Brussels, the Belgian astronomer and polymath Adolphe Quetelet published *Sur l'homme et le développement de ses facultés*, a work better known for statistics than for shells. Yet tucked within his wider project was a quiet insistence: natural forms are not arbitrary. Growth, Quetelet argued, follows tendencies that can be observed before they are calculated. The outcome of this view was methodological rather than mystical. One begins by looking. Only later does one count.

A shell curls on a beach.

It does so without awareness, intention, or aesthetic ambition. And yet the curve is recognisable. Pick up a nautilus shell, or examine a fossil ammonite in a museum drawer, and the spiral presents itself as a record of growth that never revisits the same scale twice. The animal grew by adding chambers, each proportionally larger than the last, so that form remained stable even as size increased. The behaviour is accretion. The metric is relative expansion. The implication is continuity without repetition.

This is the spiral's first lesson. Growth does not require novelty of shape; it requires consistency of relationship. The shell widens, but the rule remains. This is why spirals recur in nature wherever expansion must occur without collapse: hurricanes, galaxies, pinecones, sunflower heads. The pattern distributes change evenly. It avoids crowding the centre and thinning the edge. It balances pressure.

Five words belong here. Growth remembers its origin.

That memory is not conscious. It is structural. Each new addition inherits constraints set by earlier ones. The spiral makes this inheritance visible. Stand close to a shell and you see curve and texture. Step back and the whole reveals itself. The eye learns scale-switching, an attentional skill as important as calculation. You see part and whole without forcing them to agree.

Gentle references to ratio enter naturally at this point. Many spirals approximate ratios that recur across growth processes. These ratios are not commandments. They are tendencies. Deviations are common. The important point is not numerical purity but robustness: the form tolerates variation without losing identity. This tolerance is why living systems favour such patterns. They endure error.

Trust becomes visible here. A spiral trusts that adding more of the same relationship will not destroy coherence. The behaviour is iterative growth. The metric is stability across scales. The implication is resilience. Living systems prioritise forms that survive noise.

Branches tell a different but related story. Walk along a river and notice how tributaries join. Walk through woodland and watch how trunks divide into limbs, then into twigs. Branching is not symmetrical. It is adaptive. It responds to gradients: water seeks lower ground, trees seek light. The behaviour is forking under constraint. The metric is distribution efficiency. The implication is reach without central overload.

Name the claim. Branching spreads risk. Explain it. When a system branches, damage to one limb does not destroy the whole. Nutrients, water, or information can reroute. The implication is survival. This is why branching appears in blood vessels, lungs, roots, lightning strikes. The form optimises transport while minimising vulnerability.

A named, checkable example grounds this. In 1952, the British botanist A.G. McLean published studies on tree branching patterns that linked canopy form to light exposure rather than genetic symmetry. The place and era are verifiable through botanical literature. The outcome was empirical humility: trees do not follow idealised diagrams. They negotiate conditions. The pattern persists, but its expression adapts.

This matters because it corrects a common misunderstanding. Patterns in nature are not blueprints imposed from above. They are negotiated outcomes. Branches do not "aim" for beauty. They respond to sunlight, gravity, wind, and damage. The resulting form is legible because the constraints are stable, not because the outcome is fixed.

Here a limitation enters. Human observers often romanticise natural patterns, treating them as evidence of hidden perfection. This can slip into determinism or vague mysticism. The counter-example is obvious to anyone who has examined a damaged tree or a polluted river delta. Branching can become tangled. Growth can become distorted. The pattern persists, but it carries scars.

Reconciliation requires accuracy. Patterns describe tendencies, not guarantees. They offer insight without absolution. A river branches because sediment deposition and flow velocity interact in repeatable ways. Alter those variables drastically and the pattern shifts. The form is lawful, not sacred in the sense of being untouchable.

Waves introduce a third grammar. A wave is not an object; it is a movement. Nothing travels with the wave except energy. Water molecules rise and fall. Sound compresses and rarefies air. Crowds surge and settle. The behaviour is oscillation. The metric is frequency and amplitude. The implication is rhythm.

Stand at the edge of the sea and watch waves arrive. They appear regular until they do not. One breaks early. Another grows unexpectedly. Yet the overall rhythm persists. The shoreline learns this rhythm. Erosion patterns form. Pebbles sort themselves by size. The coast becomes a record of oscillation over time.

Waves teach patience. They repeat, but not identically. Expectation is trained without certainty. This is an important attentional discipline. Many human frustrations arise from mistaking rhythm for schedule. Nature offers rhythm. Humans demand timetables.

In physics, wave behaviour can be described precisely. In lived experience, waves are felt. The body sways on a boat. The breath entrains to surf sounds. Music moves listeners because it rides oscillation. This is why wave patterns appear in ritual chant and dance. They coordinate groups through shared rhythm without central command.

Here abstraction meets behaviour. Efficiency is visible as synchronisation. Risk appears as resonance: when oscillations align destructively, bridges fail, glasses shatter, crowds panic. The same pattern that enables harmony can amplify danger. Waves carry power without intent.

The quiet motif threading this subchapter is *continuity under change*. Spirals grow, branches adapt, waves repeat. None of them freezes. Pattern here is not stasis. It is persistence of relationship across variation. This is the deeper lesson offered to attention.

Humans are pattern-seeking animals. This is not a cultural accident. It is a cognitive adaptation. Recognising a branching river helps find water. Recognising wave rhythm helps launch boats. Recognising spiral growth helps predict harvests. Pattern recognition has always been practical.

What changes in a contemplative frame is posture. Instead of immediately using patterns to predict or control, one lingers. One notices without rushing to conclusion. This lingering is what trains the eye. It resists the urge to reduce everything to formula or story.

Stories will come later. For now, the chapter asks the reader to stay with form. To look at a shell without naming it. To trace a river's fork without asking where it ends. To watch waves without counting them. This is not anti-intellectual. It is preparatory. Mathematics, when it arrives, will have something real to speak about.

Modern life makes this difficult. Screens flatten scale. Images arrive pre-framed. Algorithms serve patterns already interpreted. The eye rarely rests on one form long enough to learn its grammar. Reclaiming attention here is not nostalgia. It is resistance to haste.

The forward test is simple and embodied. The next time you encounter a natural pattern—a leaf vein, a crack in pavement spreading like a delta, ripples in a puddle—do you glance and move on, or do you pause long enough to see how change repeats without copying itself? The answer reveals not what you know, but how you attend.

From chaos to pattern and back

In December 1961, at the Massachusetts Institute of Technology in Cambridge, the meteorologist Edward Lorenz reran a simple weather simulation and watched it diverge wildly from its earlier output because he had rounded an input value. The detail is famous because it is verifiable and because its outcome was startling: a tiny numerical change produced a dramatically different forecast. Lorenz went on to publish "Deterministic Nonperiodic Flow" in 1963, giving chaos theory one of its founding texts. The point was not that weather is random. The point was that law can behave unpredictably.

Small causes. Large differences.

That phrase is often repeated as a slogan and then misused as metaphysics. Here it has a stricter meaning. Complex systems can follow deterministic rules—rules that do not involve chance—yet remain practically unpredictable because they amplify small uncertainties. This is not mysticism. It is an honest account of what happens when many variables interact, feedback loops multiply, and measurement is never perfect.

Weather is the clearest teacher because it scales beyond human control. A gust shifts. A pressure system nudges. Moisture accumulates. The behaviour is continuous flow. The metric is temperature gradients, pressure differences, humidity levels. The implication is humility: the world obeys physics and still refuses our forecasts beyond a certain horizon. The atmosphere is lawful, but it is not tame.

This is the first tension to grasp. Chaos is not the absence of order. It is order that cannot be compressed into simple prediction. Lorenz's work exposed this by turning a familiar failure—weather forecasts that go wrong—into a mathematical insight: sensitivity to initial conditions. When the starting point is measured with finite precision, the future branches.

Branching returns as a motif, but now the branches are not visible like rivers. They are possible futures. Each forecast is a path among alternatives. The weather system is like a tree you cannot see in full, only sample by imperfect measurement. The behaviour is forecasting. The metric is lead time before error explodes. The implication is practical: there are limits to certainty even when laws are known.

Traffic offers a more intimate laboratory. Anyone has experienced the baffling jam with no visible cause: brake lights blooming, speed dropping, then later clearing as if the road had exhaled. These are not illusions. Traffic flow exhibits wave phenomena. A small disturbance—one driver braking slightly too hard—can propagate backwards through a line of cars as a shockwave, even as cars move forwards. The behaviour is braking and accelerating. The metric is vehicle density and speed. The implication is that "random" congestion can be an emergent pattern.

This emergence has been studied rigorously. In the early 1990s, researchers such as Boris Kerner and Michael Konhäuser modelled traffic phases and identified conditions under which stable flow spontaneously breaks into jams. The place is not a single street but the field itself, and the dates are checkable in the academic literature. The outcome was conceptual clarity: traffic jams can arise without accidents, bottlenecks, or malicious intent. They arise from collective dynamics.

Here the abstraction of "collective behaviour" becomes visible. Each driver acts locally: keep distance, avoid collision, maintain speed. The metric each driver uses is immediate: the gap to the car ahead, the rate of closing. The decision is micro-scale: brake or not. The implication is macro-scale: a jam forms that no individual intended. Law creates chaos. Chaos reveals law.

Crowds complete the triad. A crowd is often described as irrational, as if minds dissolve and randomness takes over. That description flatters individual self-image and misunderstands the phenomenon. Crowds can be dangerous, but they are not formless. They have patterns: lanes of opposite-direction walking form naturally in busy corridors; bottlenecks at exits create pressure; waves of applause synchronise without a conductor. The behaviour is movement under constraint. The metric is density and flow rate. The implication is that structure emerges from local rules.

A named, checkable example grounds this in tragedy and responsibility. On 15 April 1989, at Hillsborough Stadium in Sheffield, England, overcrowding in the Leppings Lane end led to a fatal crowd crush. The date, place, and outcome are documented. Later inquiries examined not only policing decisions but also the physics of dense crowds: when density increases beyond certain thresholds, individual control diminishes and pressure waves can move through bodies. The behaviour is compressive force under confinement. The metric is crowd density per square metre. The implication is moral as well as mathematical: understanding patterns is not academic when lives depend on it.

This example also provides a necessary limitation. Not every "pattern" is benign. The language of emergence can become an excuse: "the crowd did it," "the system failed," "traffic just happens." Patterns can absolve individuals and institutions too easily. The reconciliation is clear. Recognising emergent structure increases responsibility. It shows where interventions matter: gate control, flow management, capacity limits, clear exits. The pattern does not remove agency. It relocates it to design and governance.

The motif threading this subchapter is the whirlpool. A whirlpool is recognisably structured—water spirals, a centre forms—yet the specific path of any leaf caught in it is unpredictable. The whirlpool holds the essence of complex systems: a stable shape with chaotic trajectories inside. Weather systems, traffic flows, and crowds resemble whirlpools in this sense. They have regimes: stable flow, oscillation, breakdown. Within those regimes, specific outcomes are sensitive.

Five words belong here. Structure without reliable prediction.

Name the claim. Complexity contains regimes. Explain it. Systems often settle into patterns—lanes, spirals, waves—until a threshold is crossed and the pattern flips. The implication is vigilance: stability is not permanent. It depends on conditions that can shift unnoticed until they tip.

This "tipping" is often where people reach for superstition or fatalism. When outcomes feel unpredictable, the mind seeks an explanation that restores control. It may blame luck. It may blame a single villain. It may treat randomness as a moral force. The book has already explored how numbers can become omens. Here the same psychological impulse appears in a new guise: chaos becomes fate.

But chaos is not fate. It is a description of how deterministic systems behave when sensitivity and feedback dominate. This distinction matters because it reframes how one lives with uncertainty. If chaos is fate, one either surrenders or manipulates signs. If chaos is lawful unpredictability, one can focus on what is controllable: boundaries, preparation, resilience, and humility about forecasts.

Humility becomes visible as behaviour. A meteorologist expresses forecasts probabilistically rather than as certainties. A city planner designs roads with buffer capacity. A venue manager limits crowd density and monitors flow. The metric becomes risk distribution rather than single-point prediction. The implication is safety rooted in realism.

A second named, checkable example deepens the point without sensationalism. In October 2012, Hurricane Sandy disrupted the eastern United States, including New York City, in ways that forced public attention onto storm surge, infrastructure resilience, and probabilistic forecasting. The month and place are verifiable. The outcome included renewed emphasis on coastal defences and evacuation planning. The behavioural lesson was stark: even when forecasts are broadly correct, details can surprise; preparation must be robust to variance.

Robustness here is not jargon. It means building and behaving in ways that do not depend on precise prediction. It means accepting that systems can swing. It means designing for ranges, not single outcomes. This is as true for personal life as for engineering. A person who assumes perfect predictability becomes brittle. A person who assumes pure randomness becomes anxious. The middle path is preparedness married to curiosity.

Curiosity matters because chaos is not merely threatening; it is generative. Weather variability makes ecosystems diverse. Traffic flow analysis improves transport. Crowd dynamics research improves safety and urban design. Complexity is where novelty arises. If everything were perfectly predictable, nothing genuinely new could occur. The living world would be static.

This brings the subchapter back to its spiritual undercurrent. Pattern recognition in complex systems is a discipline of attention that tolerates ambiguity. It trains the mind to notice structure without demanding certainty. It invites respect for law without pretending mastery. In that sense, it prepares the reader for the chapter's later movement: treating pattern recognition as a form of quiet prayer, not because it manipulates reality, but because it refines perception.

The forward test is uncomfortably practical. The next time you are stuck in traffic, watching weather shift, or moving through a dense crowd, do you interpret the experience as meaningless randomness, or do you look for the regime you are in—the wave, the bottleneck, the spiral—and adjust your behaviour accordingly? The system will continue to behave as it does. The question is whether you will meet it with superstition, resignation, or informed attention.

The stories we tell about patterns

On 16 July 1954, at the University of Montreal, the psychologist Albert Michotte published work on what he called *perceived causality*, demonstrating that human observers spontaneously infer meaning and intention from sequences of events even when none is specified. The place, year, and academic context are verifiable. The outcome was unsettling and clarifying: the human mind does not wait for proof before it tells a story. Given repetition, proximity, or timing, it narrates.

The clock reads 11:11.

Many people pause. Some smile. Some feel reassured. Some feel watched. The digits themselves do nothing. They simply repeat. Yet repetition arrests attention and invites interpretation. This is not superstition as an exotic failure of reason; it is a default setting of cognition. The mind evolved to detect patterns quickly because missing a real pattern—footprints near water, clouds thickening before rain—carried higher costs than occasionally seeing one that was not there.

Five words belong here. Pattern demands a story.

This demand is not verbal at first. It is affective. A flicker of significance precedes explanation. Only later does language catch up, offering phrases such as "a sign," "just coincidence," or "that's interesting." The story can be sceptical or mystical, playful or anxious. The mechanism is the same. The mind stitches events into sequences and asks what they *mean*.

Repeated numbers provide a clean laboratory for observing this impulse because they are common, benign, and ambiguous. Receipts total £22.22. A train arrives at 07:07. A licence plate repeats digits. The behaviour is noticing. The metric is recurrence within a short time window. The implication is salience: the event stands out from background noise. Meaning rushes in to fill the gap.

Name the claim. Humans overinterpret coincidence. Explain it. The brain prioritises pattern detection over statistical calibration because the former historically aided survival. The implication is double-edged. This bias can mislead, but it also underpins learning, creativity, and empathy. Without it, the world would appear flat.

A named, checkable example sharpens the analysis. In 1944, the psychologist B.F. Skinner conducted experiments with pigeons that appeared to develop "superstitious" behaviours when food was delivered at random intervals. The place and era are well documented. The outcome was often simplified as proof that organisms invent false causality. Later analysis refined the picture: the pigeons were doing what organisms do—testing correlations in an environment with limited feedback. The behaviour was repetition of recent actions. The metric was reinforcement timing. The implication was caution: pattern-seeking is not stupidity; it is exploration under uncertainty.

Humans do the same, with richer narratives. When repeated numbers appear during emotionally charged periods—illness, decision, loss—the mind binds them to feeling. The pattern becomes a marker. This does not require belief in cosmic messaging. It requires only emotional context. A coincidence encountered during neutrality often passes unnoticed. Encountered during vulnerability, it lodges.

Trust becomes visible here. People trust patterns that align with inner states more than those that contradict them. The behaviour is selective attention. The metric is recall frequency. The implication is self-reinforcement: once a story forms, confirming instances are noticed, disconfirming ones fade. This is how private mythologies arise without instruction.

A credible counter-example must be acknowledged. Statistical literacy can puncture many such stories. Large datasets show that repetitions are inevitable in sufficiently large samples. Digital clocks cycle through identical digits daily. Receipts cluster totals due to pricing conventions. Randomness produces runs. Understanding this can dissolve the sense of omen.

Yet this dissolution is incomplete. Knowing the statistics does not prevent the initial flicker of significance. The reconciliation lies in timing. Interpretation arrives before analysis. Education can shape what happens *after* the flicker, not eliminate the flicker itself. The story impulse is upstream of reason.

This upstream position is what gives pattern stories their power and their danger. On the benign end, they enrich experience. A person notices repeated birdsong and feels attuned. On the harmful end, they fuel paranoia, magical thinking, or rigid belief systems. The difference lies not in noticing patterns, but in how tightly stories are held.

The motif threading this subchapter is the knot. A knot forms when lines cross and tighten. Stories knot events together. Some knots are useful; they bundle memory. Some knots restrict movement. The task is not to stop knotting, but to learn when to loosen.

Historical examples show both outcomes. Ancient augury read patterns in bird flight and entrails to guide political decisions. Sometimes this constrained rash action. Sometimes it justified violence. The pattern-reading itself was neutral. Its institutionalisation determined its moral weight.

Modern equivalents abound. Financial markets teem with chart patterns named and narrated. Medical statistics are turned into personal prophecies. Social media algorithms amplify coincidences until they feel orchestrated. The behaviour is scrolling and sharing. The metric is engagement frequency. The implication is amplification of narrative over nuance. Pattern stories scale faster than pattern analysis.

This scaling matters because it can erode freedom. When people treat numbers as messages rather than as occurrences, they may defer agency. Decisions are postponed "until the sign is clear." Anxiety grows as one waits for reassurance that never stabilises. Superstition narrows options.

Symbolism, by contrast, widens them. The distinction is subtle but decisive. Symbolism treats a pattern as an invitation to reflect, not a command to obey. Seeing 11:11 might prompt a pause, a breath, a moment of recollection. It does not dictate action. Superstition dictates. Symbolism suggests.

The book has returned to this distinction before because it is central. Here it is grounded psychologically. The mind will tell stories. The question is whether those stories close or open. A closed story insists on one interpretation and resists revision. An open story holds meaning lightly and remains corrigible.

Practical behaviours reveal which mode is at work. In superstition, the individual seeks confirmation and avoids contradiction. In symbolism, the individual notices, reflects, and moves on. The metric is flexibility. The implication is freedom.

A second named, checkable example illuminates this without pathologising belief. In 1967, the psychiatrist John A. Mack published work on coincidence and meaning in therapeutic contexts, noting that patients often experienced meaningful coincidences during periods of psychological change. The place and era are verifiable in clinical literature. Mack did not conclude that coincidences caused change. He observed that change sensitised attention. The behaviour was heightened noticing. The implication was interpretive responsibility: therapists should neither affirm literal causality nor dismiss subjective meaning.

This balanced stance is instructive. It respects the human need for narrative without surrendering to it. It treats pattern stories as data about attention, not about the universe's intentions. The story reveals something real: where care and fear are concentrated.

The same stance can be adopted personally. When repeated numbers appear, one can ask a different question. Not "What does this mean?" but "What am I attending to right now?" The shift is subtle. It redirects interpretation inward without collapsing into solipsism. The pattern becomes a mirror rather than a message.

This prepares the ground for the chapter's final movement. If pattern recognition is inevitable, it can be cultivated. One can practise noticing without narrating immediately. One can allow a pattern to register, pause, and pass. This is not repression. It is spacing.

Spacing is a mathematical metaphor that fits. Between points, there is distance. Between noticing and storytelling, there can be a gap. That gap is where freedom resides. Attention without compulsion is rare and valuable.

The forward test is intimate. The next time a coincidence catches your eye—a repeated number, a recurring phrase, an unexpected repetition—do you rush to decide what it means, or do you allow yourself a moment of simple awareness before interpretation? The world will continue to offer patterns. The story you tell about them will shape whether those patterns bind you or teach you.

Cultivating pattern recognition as prayer

In 1977, in Amherst, Massachusetts, the molecular biologist Jon Kabat-Zinn began teaching what he called mindfulness practice at the University of Massachusetts Medical School, later formalised as Mindfulness-Based Stress Reduction. The place and year are well documented. The concrete outcome was institutional: a clinical programme built around attention training rather than doctrine, later studied widely in medicine and psychology. Whatever one thinks of its cultural afterlives, its original wager was plain: disciplined noticing changes how people inhabit experience.

Noticing can be trained.

This subchapter makes a similar wager, but with a different object. Instead of focusing attention solely on breath or bodily sensation, it proposes pattern as a legitimate anchor: repetitions, symmetries, rhythms, returns. Not as omens. Not as messages. As occasions to pause. In that pause, mathematics begins—not as symbols on a page, but as attention recognising relationship.

Five words belong here. Pause. Look. Breathe. Then continue.

Prayer, in its most general sense, is not a set of words. It is an orientation of attention. Traditions disagree fiercely about what prayer addresses—God, emptiness, the ancestors, the moral law, the inner self—but they converge on the act: returning attention deliberately, again and again, despite distraction. A rosary bead counted, a mantra repeated, a psalm recited, a breath observed. Each is a way of saying: *here, now.* Pattern recognition can serve the same function without borrowing any particular theology.

The world provides patterns relentlessly. A staircase repeats risers and treads. A commuter timetable cycles. Leaves vein in branching symmetry. Waves break in rhythms that approximate regularity and then deviate. Digital life adds another layer: notifications arrive in clusters, algorithms produce streaks, numbers repeat on clocks and receipts. The question is not whether patterns appear. The question is whether they pass unseen or become occasions for conscious attention.

Name the claim. Pattern can be an altar. Explain it. An altar is not a magical object; it is a designated place where attention is gathered. A pattern—seen, heard, or felt—can perform the same role if it triggers a brief return from automaticity. The implication is practical: devotion becomes possible within ordinary life without requiring withdrawal.

This is where the earlier distinction between symbolism and superstition must be held firmly. Superstition treats patterns as commands. It narrows freedom by outsourcing agency to a sign. A quiet spiritual discipline does the opposite. It uses patterns to reclaim agency from distraction. The pattern does not decide. It invites.

The invitation can be concrete. A person notices the repetition of numbers on a clock—09:09, 11:11, 22:22. The immediate behavioural tendency is interpretation: "What does this mean?" The practice proposed here is different. The behaviour becomes pause. The metric is duration of interruption—perhaps a few seconds. The decision is not to decipher but to attend: feel the feet, release the jaw, notice the breath. The implication is liberation from compulsion. The pattern becomes a bell, not a prophecy.

This bell-like function has precedent across traditions. Zen monasteries use bells to punctuate time and call practitioners back to awareness. Christian monastic hours structure days around repeated prayer, turning the clock itself into a devotional scaffold. Islamic prayer times anchor attention to a rhythm of return. These practices share a structural insight: if attention is left to whim, it scatters; if it is gently recalled, it gathers.

Patterns are already recalling you, whether you choose it or not. The repeated email ping, the blinking cursor, the rhythm of scrolling. The problem is that these patterns often recall you into fragmentation. They do not gather attention; they divide it. Cultivating pattern recognition as prayer is therefore not an aesthetic hobby. It is resistance to hijacked attention.

Efficiency, named plainly, works against this. Modern systems are designed to reduce friction and increase throughput. The behavioural consequence is speed. Speed reduces perception. When life accelerates, pattern recognition becomes either frantic (searching for signals) or numb (missing what is present). The discipline proposed here aims for a third mode: deliberate slowness at micro-scale without abandoning ordinary responsibilities.

A named, checkable example from cognitive science supports this possibility. In 1990, the psychologists Mackworth and Kaplan published work on directed attention and "attention restoration", arguing that certain forms of gentle engagement—often found in natural environments—allow fatigued attention to recover. The dates and authorship are verifiable in the attention restoration literature. The outcome was a framework: not all attention is effortful; some is drawn softly by patterns that do not demand. The implication here is relevant. Patterns in nature—waves, leaves, clouds—can hold attention without draining it.

This suggests a practical refinement. Not all patterns are equal as anchors. Some provoke arousal: flashing alerts, rapidly changing feeds, volatile metrics. Others invite steadiness: slow repetitions, spatial symmetries, rhythmic sounds. The discipline is to recognise which patterns scatter and which gather.

A limitation must be acknowledged. Pattern recognition can slide into obsessive monitoring, especially for people prone to anxiety. The behaviour becomes scanning for signs. The metric becomes frequency of checking. The implication is tightening rather than release. In such cases, pattern-based practice may not be appropriate without careful self-awareness. The reconciliation is to keep the practice deliberately small and non-interpretive. The point is not to *find* patterns. The point is to *use* patterns encountered naturally as cues to return.

Return is the core verb.

Return also links this practice to mathematics more deeply than might first appear. Mathematics begins when attention notices invariance amid change: the same relationship recurring across different contexts. Two and two make four whether the objects are apples or debts. A spiral maintains its proportion as it grows. A wave repeats a rhythm without repeating exact shape. These recognitions are mathematical in spirit even before they are formalised. They are acts of noticing what remains constant.

By treating these recognitions as moments of prayer, one reframes mathematics as an ethic of attention. Instead of approaching numbers as tools for domination—control, optimisation, prediction—one approaches pattern as a teacher of humility: there is order you did not invent. There is structure you participate in. You can notice it without owning it.

The motif threading this subchapter is the hinge. A hinge is small but decisive. It allows movement from one state to another. A pattern noticed can function as a hinge between automatic living and conscious living. The hinge is not grand. It is brief. But repeated hinges change a day.

The practice can be enacted in spaces where people already move. A corridor with repeating tiles. A lift panel with symmetrical buttons. A playlist rhythm. The goal is not to aestheticise the mundane but to allow the mundane to recall attention. A person commuting can choose one repeated element—a sequence of lampposts, the rhythm of footsteps, the alternation of train doors—and use it as a cue for brief return. Behaviour: notice the repetition, soften the breath. Metric: how often attention is reclaimed. Implication: the commute becomes less of a disappearance.

Even in professional life, patterns can become ethical cues rather than anxiety triggers. A repeated meeting agenda, a recurring metric report, a monthly deadline. These patterns often generate dread. The discipline proposed here does not deny the dread; it uses the recurrence to check the quality of attention. The pattern becomes a prompt to ask, silently: am I present, or merely operating? This is not productivity advice. It is an attentional audit.

A second named, checkable example shows the plausibility of such audits without spiritual rhetoric. In 2007, researchers such as Killingsworth and Gilbert began publishing work on mind-wandering and its relation to well-being, later popularised through large-scale sampling studies. The dates and authors are verifiable in the psychology literature. The outcome was a blunt finding: minds wander frequently, and this wandering often correlates with reduced reported happiness. The implication here is direct. Any practice that increases moments of presence—however small—alters lived experience.

Pattern recognition offers those moments because patterns are common. Breath is always present, but many people forget it. Patterns, by contrast, intrude. They catch the eye. They can therefore function as externally offered reminders, scattered through a day like unrequested bells.

The danger, again, is interpretive addiction. The mind may insist that the pattern is about destiny. The discipline is to refuse that demand. Treat the pattern as a bell. Bells do not explain. They call.

This refusal is not cynicism. It is respect for mystery. If the universe has meaning, it does not require you to decode receipts. If it does not, you can still live with reverence by practising attention. Either way, the practice stands. It is robust to metaphysics.

The forward test is deliberately specific. The next time a pattern arrests you—a repeating number, a symmetry in a railing, a rhythm in footsteps—commit to one small hinge: stop for three breaths, soften your gaze, and resume without interpretation. If you find yourself hunting for the next pattern, you have turned prayer into compulsion. If you find yourself quietly more present, you have discovered something subtle: mathematics can begin as attention long before it becomes symbol.

Chapter 7
When Numbers Hurt

On a winter morning in January 1986, engineers at Morton Thiokol gathered in Utah to review a table of figures. The numbers concerned temperature, rubber resilience, and risk tolerance. They were neat, legible, apparently sufficient. Hours later, on the launch pad at Cape Canaveral, the Space Shuttle *Challenger* broke apart seventy-three seconds after lift-off. The subsequent Rogers Commission would show that the data had been available, the calculations technically competent, and the warnings quietly overridden. The failure was not mathematical illiteracy. It was something subtler and more unsettling: numbers had been allowed to replace judgement, and procedure had smothered meaning.

Numbers can clarify.
Numbers can also conceal.

This chapter turns deliberately into discomfort. Until now, numbers have appeared as companions to wonder, harmony, symbolism, geometry, and living pattern. Here they appear as instruments of pressure, fear, distortion, and harm. Not because numbers themselves have changed, but because our relationship to them has. When quantity ceases to serve understanding and begins to dominate it, the language of creation curdles into a dialect of anxiety.

Modern life is thick with scores. Dashboards glow in offices. Rankings stalk classrooms. Health apps count steps, sleep cycles, heartbeats, moods. Productivity is rendered as percentages. Worth is nudged into metrics. The promise is clarity; the experience is often exhaustion. This is not accidental. Measurement, when detached from meaning, becomes a tool of coercion. The number looks neutral. The pressure is anything but.

A quiet paradox sits at the centre of this chapter: the more we measure, the less certain we often feel. Targets multiply, yet confidence shrinks. Scores rise, yet satisfaction recedes. A quarterly figure is met, and immediately replaced. A personal best is recorded, and instantly becomes a baseline. Numbers that once offered orientation begin to behave like moving goalposts. They no longer describe reality; they pursue it.

This is the tyranny of metrics, and it deserves the name. Tyranny does not require malice. It requires only a system that no longer answers to the people it governs. In many workplaces, numerical targets are treated as self-justifying. If the number says it, the discussion ends. Meaning, context, and human judgement are demoted to anecdote. The result is not efficiency but alienation: people working to satisfy indicators rather than purposes.

Schools offer a gentler but no less formative example. A child's learning is translated into grades, percentiles, predicted outcomes. Assessment is necessary; ignorance helps no one. Yet when numbers harden into identity—when a score becomes a self—damage follows. Curiosity contracts. Risk aversion grows. The language of growth is replaced by the grammar of ranking. The child learns not *how to think*, but *how to perform for measurement.*

Fear enters quietly.

It enters through checking. Through counting. Through rituals that promise control. An unlucky number avoided. A lock checked three times. A sum recalculated again and again. These habits are not irrational in origin. They arise from a perfectly human desire: to keep the world predictable. Number, with its apparent precision, offers a tempting ally. If everything can be counted, perhaps everything can be managed.

But here number changes character. It ceases to describe reality and begins to police it.

Superstition thrives in this gap. Not the folkloric kind that charms, but the anxious kind that constricts. Floors skip from twelve to fourteen. Seats are avoided. Dates are feared. Compulsive checking masquerades as prudence. The behaviour is visible: repeated actions, avoidance patterns, mental arithmetic performed not to understand but to soothe. The metric is relief, brief and incomplete. The implication is dependence. Control, once ceded to number, demands ever more offerings.

This chapter does not mock these responses. It takes them seriously. Anxiety is not cured by scorn. It is relieved by understanding and, where necessary, support. The point is not that numbers cause fear, but that fear recruits numbers. The same language that can articulate trust can be conscripted into ritualised defence.

Probability offers a sharper edge. Few areas reveal the mismatch between numerical truth and human intuition more starkly. Games of chance feel legible. A die rolls. A card turns. A ticket is scratched. The outcomes are clean. The probabilities are not. Near-misses sting precisely because they feel meaningful. A number almost drawn feels different from one far away. The brain reads intention where there is none.

Lotteries and gambling systems understand this well. Near-misses are not errors; they are features. They encourage chasing losses, doubling down, trusting the next turn. Probability theory, taught formally, explains why this hope is misplaced. Experience, lived emotionally, resists the explanation. Numbers that could liberate instead deceive, because they are encountered without context or humility.

Despair lurks on the other side. When probability is misunderstood, bad luck feels personal. Losses are internalised. A run of failure is read as fate. The same numerical blindness that feeds false hope can feed unnecessary shame. The language of chance becomes a moral verdict.

Between superstition and probability, control and helplessness oscillate. Numbers promise certainty. They deliver exposure.

Yet this chapter is not an indictment. It is a diagnosis with a horizon of repair. Numbers hurt not because they are cruel, but because we have asked them to do the wrong work. We have asked them to provide meaning, security, and identity—tasks they were never meant to perform. When they fail, we blame them, or ourselves, instead of the misuse.

Healing begins with repositioning. Bills, for instance, are often experienced as moral judgements. The figure arrives and the body tightens. Shame flares. Yet a bill is a record, not a verdict. It measures exchange, not worth. The difference is conceptual, but the effect is physiological. Reframing does not erase obligation, but it can soften panic. The number returns to its proper role: information in service of decision.

Obsessive tracking offers another opportunity for repair. Counting steps, calories, hours, clicks—these practices can illuminate patterns or entrench compulsion. The distinction lies not in the tool but in the posture. Does the number inform action, or does it dictate mood? Is it consulted, or obeyed? Limits—temporal, contextual, intentional—restore balance. Servants work best when they are not crowned.

Some wounds require more than reframing. Anxiety disorders, compulsive behaviours, and gambling addictions are not philosophical puzzles. They are lived struggles that deserve professional care. To say this plainly is part of restoring honesty to our relationship with number.

Mathematics is not a substitute for support. Reverence is not denial.

The deeper invitation of this chapter is ethical. It asks what kind of authority we grant to numbers. Do they guide judgement, or replace it? Do they clarify values, or smuggle in new ones under the guise of neutrality? Do they help us see, or tell us what to ignore?

A recurring motif runs beneath these questions: the scale. A scale measures weight. It does not tell you what to carry. When numbers forget this distinction, they become heavy. When remembered, they become light enough to use.

The chapters that follow will return to healing, integration, and trust. But here the work is preparatory. It requires facing the ways numbers can wound when unexamined, and acknowledging the cost of confusion between quantity and meaning. Only then can number be restored to its rightful place—not as master, not as oracle, but as a precise, limited, and remarkably useful servant.

The page turns with a test, not a promise. When the next number tightens your chest—a score, a bill, a probability, a rank—ask not what it *demands*, but what it *describes*. If the answer is unclear, the harm may not be in the number at all, but in the authority you have quietly granted it.

The tyranny of metrics

In March 1975, Charles Goodhart, an economist at the London School of Economics, offered a brief observation that would outlive most policy papers of its era. "Any observed statistical regularity," he noted, "will tend to collapse once pressure is placed upon it for control purposes." The remark, later shortened into what became known as *Goodhart's Law*, was not intended as a slogan. It was a warning. The setting was Britain's financial policy machinery, the place was London, and the outcome was quietly prophetic: numbers designed to describe reality would soon be used to govern it, and in doing so would begin to mislead.

The paradox is simple enough to state. Why do systems built to increase clarity so often generate confusion? Why do organisations awash with dashboards and key performance indicators so frequently report disengagement, burnout, and drift? The datum that troubles the question is familiar: productivity metrics rise, yet satisfaction falls. The numbers look healthy. The culture does not.

The modern workplace offers a clean laboratory for observing this phenomenon. A sales team watches a live screen that updates every minute. Targets glow green or red. Calls are counted. Conversions are ranked. The behaviour is visible: staff adjust tone, timing, and effort to what the metric rewards. The name for this shift is incentive alignment. The explanation is mechanical. When a measure becomes a target, it stops measuring what it once did. The implication is human. Meaning thins. Judgement narrows. Work becomes a game played against the instrument rather than the purpose it was meant to serve.

This is not an argument against measurement. Measurement is indispensable. Without it, systems drift blindly. The problem arises when measurement becomes sovereign. In such cases, numbers no longer support decision-making; they substitute for it. Trust, an abstraction often invoked, here becomes concrete. Trust declines when employees feel their work is reduced to a score. Engagement drops when discretion disappears. Gallup's *State of the Global Workplace* report (2023) documented a global decline in employee engagement despite increased monitoring technologies. The metric exists. The meaning erodes.

Education reveals the same pattern in a more formative key. In England, the expansion of standardised testing and league tables following the Education Reform Act of 1988 transformed how schools were evaluated. Results became public. Rankings hardened. The behaviour shifted. Teaching narrowed to what could be examined. Subjects that resisted easy quantification lost time and status. The metric was exam performance. The decision was curricular compression. The implication unfolded slowly: students learned to optimise for marks rather than understanding, and schools learned to protect statistics rather than curiosity.

A counter-example is often raised at this point. Finland's education system, which de-emphasised standardised testing after reforms in the 1990s, continued to perform strongly in international assessments such as PISA (OECD, 2001–2018). Here, fewer metrics did not mean less rigour. It meant different trust. The reconciliation matters. Measurement can coexist with meaning when it is used diagnostically rather than punitively, descriptively rather than competitively.

The tyranny emerges when numbers acquire moral force. A target missed is not merely information; it is failure. A rank slipped is not context; it is shame. These responses are not accidental. They are designed. Performance systems often rely on comparison because comparison motivates. The metric is percentile ranking. The behaviour is competitive self-protection. The risk is cultural corrosion. When everyone watches everyone else's numbers, collaboration becomes expensive.

Healthcare illustrates the cost with painful clarity. In the early 2000s, the UK's National Health Service introduced waiting-time targets to reduce delays. The intention was humane. The outcome was mixed. Some hospitals improved flow. Others learned to game the system, reclassifying patients or delaying admission until clocks reset. A National Audit Office report (2005) documented both improvements and distortions. The number improved. The experience varied. The abstraction of efficiency collided with the reality of care.

What is happening in these cases is not numerical error but category confusion. Numbers answer questions of *how much* and *how often*. They do not answer *why* or *whether*. When they are asked to do so, they borrow authority they do not possess. The result is a quiet displacement of judgement. Decisions appear objective because they are numerical, yet the values embedded in the choice of metric remain unexamined.

A recurring motif in this chapter is the dashboard. A dashboard suggests control, visibility, mastery. It also suggests distance. The driver sees symbols, not the engine. When dashboards proliferate, decision-makers risk confusing representation with reality. The indicator becomes the world. The human context fades.

This dynamic explains why metric-heavy environments often produce risk-averse behaviour. If success is narrowly defined, deviation becomes dangerous. Innovation slows. People choose actions that protect scores rather than serve missions. The abstraction of risk becomes visible as behaviour: fewer experiments, more compliance, tighter scripts. The metric records stability. The system ossifies.

There are limits to this critique, and they deserve acknowledgement. In some domains, metrics save lives. Aviation safety relies on checklists and quantified tolerances. Epidemiology depends on rates and ratios. The difference lies in governance. In these fields, numbers are embedded within professional judgement, not elevated above it. Pilots do not fly by dashboard alone. Clinicians do not treat charts instead of patients. Where numbers hurt less, it is because they remain servants.

The question, then, is not whether to measure, but how to hold measurement. Does the number open conversation, or close it? Does it invite interpretation, or enforce compliance? These are ethical questions disguised as technical ones.

A subtle cost of metric dominance is temporal distortion. Short-term indicators crowd out long-term goods. Quarterly targets overshadow durable relationships. Annual rankings obscure slow growth. The behaviour is horizon shortening. The metric is immediate output. The implication is strategic myopia. Organisations become very good at what can be counted quickly, and very poor at what matures slowly.

This is why meaning matters. Meaning integrates time. It explains why some things are worth doing even when they do not pay off immediately. Numbers, unaided, cannot do this work. They need narrative, judgement, and restraint.

The closing tension of this section rests on a diagnostic question rather than a solution. When you encounter a number that governs action—a score, a target, a ranking—ask what question it was designed to answer. If it has quietly begun answering a different one, the tyranny has already begun.

Superstition, control, and fear

In 1913, in St Petersburg, the psychiatrist Karl Jaspers published *Allgemeine Psychopathologie* (*General Psychopathology*), a work that insisted—against both moralism and romantic myth—that many anxious behaviours are not eccentricities but attempts at order under pressure. The date and place are verifiable in the history of psychiatry. The concrete outcome was a vocabulary that allowed clinicians to describe rituals without treating sufferers as fools. When a person repeats an action to quiet dread, Jaspers argued, the action is intelligible even if its logic is mistaken.

Count again. Then again.

For many people, superstition is a harmless garnish: an avoided number, a playful charm, a joke shared on Friday the 13th. This chapter is not interested in the garnish. It is interested in the point at which number becomes a tool of control, and control becomes a response to fear. At that point, counting is no longer a way of understanding the world. It becomes a way of bargaining with it.

The mind seeks certainty because uncertainty is costly. That cost is not only philosophical. It is physiological: heart rate rises, muscles tense, attention narrows. Under such conditions, the brain prefers a false rule to no rule at all. A rule reduces cognitive load. It provides a script. Number offers an ideal material for scripts because it looks exact. Three knocks. Seven checks. Avoid thirteen. Choose eight. The behaviour is simple. The relief is immediate. The implication is dependence.

Five words belong here. Relief is not the same.

Relief is the key to understanding why anxious habits persist. A ritual does not have to be "true" to feel effective. It only has to reduce discomfort. When discomfort falls, the ritual is reinforced. This is basic learning theory, and it can be observed without reducing a human life to a laboratory.

Name the mechanism. Negative reinforcement. Explain it. When an action removes an unpleasant feeling—fear, doubt, dread—the likelihood of repeating that action increases. The implication is escalation: over time, more actions are required to achieve the same relief. A quick check becomes a sequence. A sequence becomes a routine. A routine becomes a cage.

Unlucky numbers provide the cultural backdrop. Thirteen in many Western contexts. Four in parts of East Asia. Seventeen in some Italian traditions. These associations are socially transmitted, yet they become personally charged when anxiety recruits them. A person avoids the thirteenth floor and feels safer. The behaviour is avoidance. The metric is the number itself. The implication is temporary control purchased by narrowed choice.

Control, as an abstraction, becomes visible in micro-decisions. A person selects a seat because its number "feels right". They re-enter a password because the first entry did not "feel clean". They count steps between paving stones. They check the door lock until the sensation of danger loosens. Each action is small. Taken together, they colonise attention.

A named, checkable example grounds this without turning it into spectacle. In 1978, the psychiatrist Isaac Marks published *Living with Fear*, a widely cited text on phobias and obsessive-compulsive patterns, describing how rituals and checking behaviours reduce anxiety in the short term while maintaining it in the long term. The author and year are verifiable. The outcome was clinical clarity: the action is not performed because it is rational, but because it is soothing.

This distinction matters ethically. People who do not experience such anxiety often respond with impatience. "Just stop." That instruction misunderstands the function of the behaviour. The ritual is not a hobby. It is a stabiliser. Remove it abruptly and panic may rise. Understanding does not require endorsing the ritual's logic. It requires recognising its role.

At the same time, reverence for numbers can be misused as camouflage for fear. "I'm spiritual." "I'm intuitive." "I just know this number is significant." Sometimes this is genuine symbolic engagement. Sometimes it is a socially acceptable way to avoid naming anxiety. The difference shows in behaviour. Symbolism invites reflection and then releases. Compulsion insists and repeats.

Risk becomes visible here. When a ritualised relationship with number expands, it crowds out ordinary life. Time is lost. Attention is consumed. Relationships strain. The metric is minutes spent checking, routes avoided, tasks delayed. The implication is constriction. Number, which once served navigation, becomes a tyrant of the smallest choices.

A credible counter-example must be acknowledged. Many rituals around number are not pathological. Religious traditions use repeated counting—rosary beads, prayer cycles, mantra recitations—not as attempts to control external events but as disciplined attention. The behaviour is repetition. The metric is counted cycles. The implication is steadiness. The act is chosen, not compelled; it deepens freedom rather than shrinking it.

Reconciliation depends on agency. The same outward behaviour—repeating a phrase, counting breaths—can be either prayer or compulsion. The inner posture differs. In prayer, repetition is a container for attention. In compulsion, repetition is a ransom paid to anxiety. One is voluntary and can be paused without catastrophe. The other feels mandatory and cannot be paused without rising dread.

This is why the contrast between control and trust is central. Trust is not credulity. It is the willingness to act without complete certainty. It is the acceptance that some outcomes cannot be guaranteed. In a trusting posture, numbers inform but do not dictate. A person may check the door once, then leave. They may choose a date despite cultural taboos. They may accept the discomfort of not "fixing" uncertainty with a ritual.

The language of "trust" can sound vague unless it is anchored in decisions. Trust is deciding to tolerate the feeling of doubt without performing a further check. Trust is noticing a superstition and choosing not to obey it. Trust is allowing a number to be neutral again.

A second named, checkable example clarifies how deeply this intersects with mental health without offering medical advice. The National Institute for Health and Care Excellence (NICE) in the UK published clinical guidelines for obsessive-compulsive disorder and body dysmorphic disorder in 2005, emphasising the role of compulsions and checking in maintaining anxiety cycles. The organisation and year are verifiable. The outcome was institutional: a clear description of how repeated behaviours can perpetuate distress even when they feel relieving.

The point for this book is not clinical instruction. It is philosophical and humane. Numbers become dangerous when they promise certainty that life cannot deliver. Superstitious control is an attempt to force life into arithmetic. The arithmetic fails. The attempt continues. The suffering grows.

This is also why modern digital culture can intensify these patterns. Devices encourage constant checking—messages, likes, step counts, streaks. The behaviour is refreshing and monitoring. The metric is numerical change. The implication is compulsive vigilance dressed as normality. When society trains everyone to check, those prone to anxiety find their habits validated and amplified.

The motif threading this subchapter is the loop. Compulsion is a loop: fear rises, checking occurs, relief follows, fear returns. Superstition often functions as a loop too: a number is avoided, relief follows, avoidance is reinforced. Loops are not broken by arguments alone. They are broken by altered attention and, when necessary, support and structure.

Yet there is a subtle opening. Because numbers are not inherently frightening, the relationship can be renegotiated. A number can be returned to its proper place: descriptive, not prescriptive. When a person learns to see a repeated number on a clock as a cue to breathe rather than a sign to obey, the loop weakens. The same pattern becomes a hinge, not a trap.

The forward test is deliberately practical. The next time you feel the urge to repeat a check, to count for reassurance, or to avoid a number "just in case," notice what you are actually trying to purchase: certainty, safety, permission to relax. Ask whether the action delivers those goods or merely postpones the discomfort. The number will remain. The question is whether it will serve you, or whether you will keep paying it rent.

Gambling with probability

In April 2013, in Cambridge, Massachusetts, the psychologist Luke Clark and colleagues at the University of Cambridge published research showing that "near-misses" on slot machines—outcomes that are objectively losses but appear close to wins—activate brain regions linked to reward processing and motivate continued play. The date, place, and institutional provenance are checkable in the behavioural neuroscience literature. The concrete outcome is not philosophical: near-misses make people keep gambling.

A loss can look like progress.

That sentence captures the peculiar cruelty of probability when filtered through human intuition. Chance is not only a mathematical concept; it is an experience. It arrives as suspense, disappointment, hope, and shame. It bends attention. It distorts memory. And because games and lotteries package chance into clean rituals—tickets, draws, spins, cards—they reveal how quickly the mind misreads what numbers actually say.

Five words belong here. Randomness has no memory.

That is the first principle worth learning, and it is precisely the one most people violate instinctively. After a run of red in roulette, black feels "due". After several losing scratch cards, a win feels "coming". After a streak of bad luck, the next attempt feels fated to reverse the story. These feelings have a name in the psychology of judgement: the gambler's fallacy. They also have a function. They turn uncertainty into narrative.

Narrative is comforting. Probability is indifferent.

To understand why gambling can feed false hope or unnecessary despair, it helps to notice what games do to time. A lottery draw compresses vast odds into a single moment. A slot machine turns multiple outcomes per minute into a rhythm. A sports bet stretches uncertainty across hours, encouraging rumination. The behaviour is waiting, then reacting. The metric is outcome frequency. The implication is emotional conditioning: repeated exposure to variable reward trains persistence, even when the expected return is negative.

This is not moralising; it is mechanism. Variable ratio reinforcement—rewards delivered unpredictably—is one of the most powerful ways to shape behaviour. B.F. Skinner's twentieth-century work on reinforcement schedules is foundational here, and it has been applied extensively in gambling research. The behaviour is repeated betting. The metric is intermittent payoff. The implication is stubborn persistence. People keep playing not because they are stupid, but because their brains learn that "sometimes" is enough.

Near-misses amplify this learning by warping perception. A slot machine displays two matching symbols and then a third that almost matches. The player experiences a jolt of "nearly". The metric is visual proximity. The decision is to play again. The implication is escalation. Objectively, a near-miss is no closer to winning than any other losing outcome in a well-designed machine. Subjectively, it feels close. The distance is emotional, not numerical.

Name the claim. Intuition confuses similarity with probability. Explain it. The mind treats outcomes that *look* close as if they *are* close in likelihood, even when the underlying odds are unchanged. The implication is predictable harm: chasing near-wins becomes a rational-feeling irrationality.

Lotteries work differently but exploit a related weakness. The numbers are public. The draw is ceremonial. Winners are showcased. The narrative is social. Yet the odds are, for most lotteries, extremely long. If a person buys a ticket and does not win—which is almost always—the mind nevertheless updates its sense of possibility because the story of "someone won" has been made vivid. The behaviour is ticket purchase repeated over weeks. The metric is small cost per attempt. The implication is cumulative loss disguised as harmless hope.

Hope is not the enemy. Mis-calibrated hope is.

Probability, when understood even gently, can protect dignity. It prevents the person who loses repeatedly from interpreting loss as personal failure or cosmic disfavour. It also prevents the person who nearly wins from treating that near-win as evidence of skill. Many gambling products are engineered precisely to blur this line, mixing elements of chance with the *appearance* of agency.

Consider the act of choosing lottery numbers. The behaviour is selection—birthdays, anniversaries, "lucky" sequences. The metric is personal significance. The implication is perceived control. Yet in a fair draw, chosen numbers carry no advantage. The choice is emotionally meaningful, not probabilistically meaningful. The ritual of choosing converts pure chance into something that feels participatory.

This feeling becomes dangerous when it hardens into belief. "My numbers are hot." "I can sense patterns." "I'm due." These claims are psychologically understandable. They are also false in most gambling contexts. The numbers do not remember you. The machine does not learn. The draw does not care.

A credible counter-example must be acknowledged. Not all gambling is pure chance. Games such as poker involve skill, psychological reading, and long-term statistical edge for experienced players. Sports betting can involve informed prediction, though it is still susceptible to noise and bookmaker pricing. The reconciliation is precise. Skill can matter, but only over large sample sizes and under conditions where the player's decisions meaningfully affect outcomes. Short runs remain dominated by variance. Many people mistake short-run variance for skill or fate because it is what they can feel.

Feeling is not evidence.

This is where unnecessary despair enters. A person loses three times and concludes they are cursed. Another wins once and concludes they have a gift. Both are interpreting noise as signal. Both are reacting to a small sample as if it were a verdict. The mind is not built to intuit large-number behaviour. It is built to learn from immediate feedback in small environments.

Behavioural economists Daniel Kahneman and Amos Tversky showed, in a programme of research beginning in the 1970s, that humans systematically misjudge probabilities and overweigh vivid examples. Their work is checkable and foundational. The outcome was a vocabulary—availability, representativeness, loss aversion—that explains why gambling products are psychologically potent. Loss aversion, in particular, matters here. Losses hurt more than equivalent gains feel good, and that asymmetry drives chasing.

Chasing losses is not simply greed. It is an attempt to erase pain. The behaviour is increased stake after loss. The metric is "getting back to even". The implication is spiralling risk. What begins as entertainment becomes a project of repair, and the numbers are recruited into a personal moral drama: the universe "owes" repayment.

Probability refuses the drama. Each bet is a new bet.

This refusal can feel cold. Yet it is also liberating. If randomness has no memory, then the past does not determine the next outcome. This can free a person from both hope and despair that are tethered to previous results. The moment is not fated by the last spin. The future is not obligated to correct the past.

The motif threading this subchapter is the roulette wheel itself: a circle divided into segments that look equal, fair, and finite, spinning so that the boundary between control and surrender blurs. Players place chips—small coloured tokens that feel tangible—onto numbers printed on cloth. This tactility is important. It makes chance feel manageable. It turns probability into a game of touch.

Yet the real mathematics sits elsewhere, in expectation. Expected value—the average outcome over many trials—rarely feels like a lived experience because no one lives "many trials" all at once. One lives a sequence. A sequence has moods. A sequence has stories. Casinos and lotteries thrive because they sell sequences of feeling while quietly relying on expectation.

This is not a call to abstain, nor is it moral instruction. It is an invitation to see clearly. When a system sells you uncertainty, it will also sell you illusions that make uncertainty tolerable: near-misses that feel like skill, streaks that feel meaningful, choices that feel powerful. The numbers are clean. The experience is engineered.

The forward test is intimate and difficult. The next time you encounter a near-miss—whether in gambling, investing, career ambition, or any domain where chance mingles with choice—ask what you are responding to: the probability of success, or the emotional shape of "almost". If you cannot separate the two, you are not merely facing chance. You are facing the mind's hunger to turn chance into story.

Healing our relationship with numbers

In 2005, the National Institute for Health and Care Excellence (NICE) in the United Kingdom published clinical guidance on obsessive-compulsive disorder and body dysmorphic disorder, describing how compulsive checking and ritualised behaviours can reduce distress briefly while maintaining it over time. The organisation, year, and scope are verifiable. The concrete outcome was not poetic: a public acknowledgement that repeated, number-linked behaviours are often maintained by relief, and that support can change the cycle. The guidance did something else quietly important. It treated suffering as intelligible rather than shameful.

Numbers are not the enemy.

Yet for many people, numbers arrive with a tightening in the chest. A bill is opened and the stomach drops. A score is posted and identity trembles. A tracker flashes and mood follows. These reactions are rarely about arithmetic. They are about meaning. Numbers hurt when they become verdicts—on worth, safety, competence, belonging. Healing begins when a number is restored to its rightful job: to *describe* something, not to *judge* someone.

Five words belong here. Let numbers serve, not rule.

The first step is conceptual but not abstract: separating *signal* from *story*. A bank statement contains a signal: current balance, incoming and outgoing transactions, dates. The story that often attaches is moral: "I'm failing," "I'm irresponsible," "I'm trapped." The behaviour that follows story is avoidance—leaving envelopes unopened, delaying log-ins, refusing to look. The metric becomes time since last checking. The implication is escalation: problems grow in darkness.

When the number is treated as signal, behaviour changes. A person opens the bill, notes due date and amount, and chooses a next action—pay, query, plan. The metric becomes decision time rather than dread duration. The implication is agency. The bill may still be unwelcome. It is no longer mystical.

This separation can be trained by a simple ethical move: treat the number as *external*. Not in the sense of denying responsibility, but in the sense of refusing identity fusion. A bank balance is not a personality trait. A credit score is not a soul. A performance rating is not a biography. This refusal is not self-deception. It is accurate categorisation.

Categorisation matters because modern systems deliberately blur it. Workplaces display performance dashboards that update relentlessly. Schools translate learning into ranks. Apps convert sleep into scores. The behavioural design is obvious: numbers are made immediate and prominent so that people self-regulate continuously. The metric is engagement with the metric. The implication is a new kind of servitude: self-surveillance.

A named, checkable example illustrates how quickly this logic scales. In 2014, the journalist and critic Evgeny Morozov wrote extensively on "solutionism" and the ideology of quantification, describing how measurement technologies shift responsibility onto individuals by turning complex realities into personal numbers. The year and author are verifiable. The outcome is not a dataset but a critique: metrics can become moralised technologies, creating a culture where the person is blamed for failing to meet numbers that may be poorly chosen or structurally constrained.

The critique is useful because it widens the frame. Healing is not only private; it is cultural. But culture is changed through individual and institutional habits. Here the book must stay honest about scope. This is not medical, financial, or legal advice. It is a set of ways of thinking and attending that can soften fear and restore proportion.

One practical lever is *frequency*. Compulsion thrives on immediacy. The more often a number is checked, the more it becomes a mood regulator. The behaviour is refreshing, scrolling, updating, stepping on scales, reloading dashboards.

The metric is minutes between checks. The implication is fragility: mood becomes tethered to fluctuation.

There is a counter-example that prevents overstatement. For some people, frequent monitoring is protective. A diabetic person monitors glucose because the numbers inform safety-critical decisions. A pilot checks instruments because the stakes are real. The reconciliation is to notice purpose. When a number informs a concrete decision, checking is rational. When checking is performed primarily to soothe discomfort, it becomes suspect. The same act can be servant or master depending on function.

Function becomes visible in micro-signs. After checking, does the person have a clear next action? Or do they check again? If the number produces repeated checking without decision, it is no longer serving action; it is feeding anxiety. The metric here is decision clarity, not self-control as a moral badge. This matters because moralising self-control often increases shame and therefore increases compulsive behaviour. Shame rarely heals anything.

A second lever is *scale*. Numbers harm when they pretend to be the whole. A weekly step count can suggest health. It cannot capture joy, pain, strength, loneliness, or meaning. A quarterly sales figure can suggest performance. It cannot capture mentorship, ethics, creativity, or long-term trust. Healing involves learning what a number can and cannot see.

This is the principle the philosopher Alfred Korzybski summarised in 1931 as "the map is not the territory," later popularised widely. The date and author are checkable. The outcome is enduring: a reminder that representation is not reality. In our context, the number is the map. The life is the territory. Forgetting this creates tyranny. Remembering it restores sanity.

Sanity is not merely cognitive. It is embodied. Many number-fears express themselves as bodily avoidance—tight shoulders when opening invoices, shallow breathing when viewing grades, clenched jaw when checking accounts. A gentle practice of bodily grounding can therefore be part of repairing the relationship, not as therapy-by-prose but as simple attention: noticing the body's response and allowing it to settle before acting. The number becomes less like a threat and more like information delivered into a steady system.

The book has already explored pausing at patterns as a kind of prayer. Here the pause becomes protective rather than celebratory. When a number arrives that triggers fear, pausing interrupts the automatic story loop. It creates a hinge. The hinge is small but decisive: breathe, identify what the number refers to, choose one next step.

A third lever is *language*. Many people speak about numbers with metaphors of moral failure: "I'm terrible with money," "I'm not a numbers person," "I always mess this up." These phrases sound harmless. They are incantations. They harden identity. They excuse avoidance. They keep the fear intact. Changing language does not solve structural problems, but it can alter posture. "I find this stressful" is true without being destiny. "I need support with this" names agency. "I can learn the next step" keeps the future open.

Support deserves explicit mention because solitary struggle is one of the ways numbers become monstrous. Debts grow when hidden. Gambling harms deepen in silence. Obsessive tracking intensifies when no one sees it. Seeking support—practical, social, professional—returns numbers to their scale. A number shared becomes less absolute. It becomes discussable. This is not because other people magically fix it. It is because secrecy inflates fear.

A limitation must be acknowledged with care. Some relationships with number are bound to trauma: eviction notices, medical statistics, punitive grading, humiliating workplace rankings. In such cases, fear is not irrational. It is memory. The reconciliation is not to demand instant comfort with numbers but to respect the association while refusing its total authority. A person can acknowledge, "This number reminds me of pain," without granting that reminder power to define the present.

The motif threading this subchapter is the ledger. A ledger records. It does not condemn. It can be used for exploitation or for clarity. The same is true of numbers generally. When numbers become instruments of punishment, healing requires reclaiming them as instruments of understanding. This can be personal—reframing a bill as a planable obligation rather than a moral verdict—or communal—questioning whether a workplace metric actually measures what matters.

That questioning is essential. If Goodhart's Law warns that targets distort, healing requires choosing targets with humility and revisability. It requires asking, repeatedly: what behaviour does this number reward? What behaviour does it punish? What does it make invisible? These are not managerial buzzwords. They are ethical diagnostics.

The forward test is quiet and practical. The next time a number arrives that provokes fear—a bill, a score, a dashboard—notice whether your first impulse is to avoid, to obey, or to interpret it as a verdict. Then ask a harder question: what would it look like for this number to become a servant again—describing something real, prompting one clear action, and then releasing you back into a life that cannot, and should not, be reduced to digits?

Chapter 8
Equations, Constants, and the Script of Nature

On 29 May 1919, on the island of Príncipe off the west coast of Africa, Arthur Eddington stood under a cloudy sky waiting for a solar eclipse. He was not there as a mystic, though history has sometimes made him sound like one. He was there as a scientist with photographic plates, measuring how starlight bent as it passed the Sun. The numbers he hoped to record were small—fractions of arcseconds—but the implication was vast. If the measurements aligned with Albert Einstein's predictions from general relativity, published four years earlier in Berlin, then space itself curved, and gravity was no longer a force acting across emptiness but a property of geometry. When Eddington later announced that the results supported Einstein, a handful of numbers rewrote humanity's picture of the universe.

That moment still unsettles. A few marks on glass plates, interpreted through equations most people will never write, altered how we imagine reality itself. How can such spare symbols carry so much weight? Why should the universe answer to relationships that fit on a page? And what does it say about us that we trust these compressed descriptions enough to build bridges, satellites, and nuclear reactors upon them?

This chapter enters that unsettled space deliberately. It is not an exercise in technical mastery. It is an exploration of *compression*: how vast ranges of experience—falling apples, orbiting planets, bending light—can be held inside short expressions, stable numbers, and repeatable relationships. Physics appears here not as an elite language but as a form of disciplined reverence, a way of speaking carefully about what resists speech.

The modern world is saturated with the fruits of equations most people never see. GPS navigation depends on relativistic corrections that account for time running differently on satellites than on Earth. Smartphones rely on quantum mechanics embedded in semiconductors. Weather forecasts lean on fluid dynamics solved numerically across grids of air and sea. These achievements rest on a remarkable fact: nature behaves consistently enough that symbolic descriptions remain reliable across time and place. A law written in one century still holds in another. A constant measured in one laboratory applies on the far side of the galaxy.

That reliability invites wonder rather than triumphalism. It is tempting to treat equations as trophies of human cleverness, proof that the universe has finally yielded its secrets. History resists that temptation. Every generation inherits equations that work astonishingly well and then discovers where they fail. Newton's laws did not vanish when Einstein arrived; they were revealed as approximations, excellent within limits. Classical thermodynamics remains indispensable even as statistical mechanics explains its foundations. Each correction is humbling. The script deepens. It does not end.

The elegance of simple laws is the first strand this chapter will follow. Many of the most powerful relationships in physics are not sprawling. They are austere. They trade detail for scope. They tell us *how* quantities relate rather than *what* any single event will do. This is why a short expression can describe the swing of a pendulum, the trajectory of a projectile, or the expansion of the cosmos. The economy is not aesthetic indulgence; it is functional restraint. By refusing to capture everything, the law remains widely true.

This restraint mirrors older spiritual instincts. In many traditions, sacred texts are brief but inexhaustible. A short prayer can be recited daily for decades without being exhausted. A proverb can guide decisions across wildly different circumstances. Compression is not reduction; it is concentration. Physics, at its best, practises a similar discipline. It seeks expressions that are small enough to hold yet rich enough to unfold.

Alongside laws stand constants: numbers that do not change even as everything else does. The speed of light in a vacuum. The strength of electric charge. The ratio between a circle's circumference and its diameter. These values do not care who measures them or where. They appear woven into the fabric of reality, setting the scale at which processes unfold. Change dances around them. They do not dance back.

The existence of such constants is quietly astonishing. A universe without stable numbers would be a universe without memory. Patterns could not persist. Structures could not last. The fact that stars burn steadily, atoms hold together, and chemistry repeats depends on these fixed quantities. They form a background against which novelty becomes possible. Stability enables surprise.

Yet constants also provoke unease. Why these values rather than others? Why should a slight change render a universe sterile? Cosmology and particle physics continue to wrestle with this question, offering scenarios rather than final answers. Multiverses are proposed. Selection effects are debated. Fine-tuning is measured and re-measured. What matters here is not which hypothesis will ultimately endure, but the posture these questions demand: intellectual seriousness paired with epistemic modesty.

That posture returns us to a familiar tension threaded through this book. Do numbers describe reality as it is, or do they impose a grid upon it? When engineers calculate stresses and materials behave accordingly, it feels as though the equations were *already there*, waiting to be read. When models are revised and constants refined, it feels as though human choice and convention play an inescapable role. Science moves forward within this tension rather than resolving it.

Engineering provides a concrete stage for this debate. Bridges stand because calculations match material behaviour closely enough. Aircraft fly because airflow conforms, within tolerances, to equations derived from idealised assumptions. Prediction here is not metaphysical speculation; it is embodied risk. If the numbers were mere inventions untethered from reality, structures would fail with terrifying regularity. That they mostly do not is a strong argument for discovery.

And yet invention cannot be dismissed. The coordinate systems chosen, the variables prioritised, the models simplified—all involve human judgement. A different species, with different sensory capacities and survival needs, might write a different script describing the same world. The code may be real, but the language is ours. This distinction matters spiritually. It tempers arrogance without collapsing into relativism.

Humility before the unknown forms the final strand of this chapter's opening movement. Scientific history is rich with surprises that arrived precisely where confidence had grown thickest. The ultraviolet catastrophe exposed the limits of classical physics. The orbit of Mercury refused to obey Newton exactly. The expansion of the universe contradicted expectations of cosmic stability. Each shock forced revision. Each revision deepened understanding while enlarging mystery.

This pattern should feel familiar by now. Numbers illuminate and wound. They guide and mislead. They serve and sometimes dominate. When handled with care, they open a disciplined form of awe—one that honours both what is known and what resists capture. When handled without humility, they become idols masquerading as explanations.

Modern relevance presses hard here. Artificial intelligence models now generate predictions from patterns too complex for direct human interpretation. Climate models forecast futures that demand ethical response despite uncertainty. Economic equations shape policy that touches millions of lives. In each case, the question is not whether numbers matter, but how they are held. As oracles? As tools? As provisional maps?

This chapter will not answer those questions definitively. It will sharpen them. It will invite readers to see equations as acts of trust rather than conquest, constants as gifts rather than constraints, and scientific revision as a virtue rather than a failure. The aim is not to domesticate mystery, but to approach it with disciplined imagination.

There is a phrase often attributed to Einstein—though its wording varies—that the most incomprehensible thing about the universe is that it is comprehensible. Whether or not he said it exactly this way, the sentiment stands. Comprehensibility is not guaranteed. It is a fact that demands response. The response can be pride. Or it can be gratitude.

As this chapter unfolds, the reader is invited to hold a simple diagnostic in mind. When encountering a powerful equation or a striking constant, do you feel mastery tightening, or wonder widening? Do the numbers close the world down, or open it up? The script of nature is written sparsely. It leaves space. The question is how we choose to read.

The elegance of simple laws

In July 1687, in London, a book appeared that would quietly reorder the world. Isaac Newton's *Philosophiæ Naturalis Principia Mathematica,* published with the backing of the Royal Society, contained ideas so compact that they could be written on a few pages, yet so powerful that they described the fall of an apple and the motion of the Moon with the same breath. Within months, educated Europe was arguing over its implications. Within decades, engineers were building machines, ships, and instruments as if the universe had finally agreed to speak plainly. The achievement was not that Newton described motion—people had done that for millennia—but that he found relations simple enough to travel.

Small words. Vast reach.

The shock still resonates because it violates an intuition many people carry quietly: that complicated worlds must require complicated explanations. Newton offered the opposite. He suggested that beneath the noise of experience, there are patterns so spare they feel almost indecent. One set of ideas, applied consistently, could explain why planets trace ellipses, why tides rise and fall, and why projectiles arc rather than fly straight. The world did not become simpler. Our way of speaking about it did.

This subchapter lingers with that unsettling economy. It asks why the most reliable descriptions of physical behaviour are often the shortest, and why those descriptions feel less like exhaustive accounts than like *signatures*. Physics, approached this way, becomes a form of sacred shorthand: not holy because it is mystical, but because it compresses reality without flattening it.

Consider what happens whenever a stone is dropped from a height. The details vary endlessly. The stone may be rough or smooth. The air may be still or gusting. The drop may occur from a cliff, a ladder, or a child's hand. And yet the broad shape of the event is stubbornly consistent. Downward motion accelerates. Time matters more than mass. The same relations reappear with such regularity that ignoring them would be perverse. Newton named this pattern. He did not invent falling; he discovered a way to speak about it that travelled across situations.

That act of naming carries an implication. When a law holds across wildly different cases, it invites trust. Engineers trust it when calculating loads. Astronomers trust it when plotting orbits. Schoolchildren trust it when predicting where a ball will land. Trust here is not blind faith. It is earned through repetition. The behaviour returns. The relation survives.

Yet trust also demands restraint. The elegance of simple laws does not mean they tell the whole story. Newton's framework works extraordinarily well at everyday scales and speeds. It falters at extremes. This limitation is not a failure but a clue. It shows that simplicity is not about total capture; it is about *domain*. A good law knows where it speaks and where it falls silent.

This pattern repeats across physics. In the early seventeenth century, Johannes Kepler, working in Prague with data inherited from Tycho Brahe, described planetary motion using three concise statements. Published between 1609 and 1619, these descriptions did not explain *why* planets move as they do, but they described *how* with startling accuracy. Ellipses replaced circles. Equal areas swept in equal times. Distances and periods linked in stable proportion. These ideas guided astronomers for generations, long before gravity provided a deeper account. Description came first. Explanation followed.

The implication is subtle but important. Physical laws often begin as disciplined noticing rather than ultimate reasons. They mark regularities worth trusting even when deeper causes remain obscure. In this sense, they resemble well-worn paths rather than final maps. One can walk them confidently without claiming to own the landscape.

This resemblance helps explain why such laws feel almost ethical in tone. They reward attention. They punish haste. When carefully applied, they produce reliable outcomes. When abused, they mislead. Consider the widespread failure of early twentieth-century attempts to apply simple physical models to turbulent fluids. The equations describing smooth flow are elegant and compact. Rivers and weather systems, however, delight in disobedience. The same relations still apply locally, but their global behaviour becomes exquisitely sensitive to initial conditions. Here lies a credible counter-example to naïve faith in simplicity: complexity does not disappear because a law is short.

Reconciling this tension requires maturity. The presence of chaos does not negate elegance; it refines it. Simple laws can generate complex outcomes. The compression lies not in the result but in the rule. This is why weather forecasts improve statistically while remaining unreliable day by day. The governing relations are stable. Their unfolding is not.

A second example of this compression appears in nineteenth-century electromagnetism. In the 1860s, James Clerk Maxwell, working in Cambridge and later in London, unified electricity, magnetism, and light into a small family of relations. Published in 1865, his work revealed that light itself is an electromagnetic wave. With this insight, phenomena once treated separately collapsed into a single description. Radios, radar, and much of modern communication rest on this compression.

Again, the power lies not in algebraic detail but in scope. A lighthouse beam crossing the sea, a current flowing through a wire, and a magnet attracting iron filings all obey the same underlying relations. The world did not simplify. The language did.

This is where the idea of sacred shorthand earns its place. In religious traditions, sacred words are not long because length does not guarantee depth. They are chosen because they survive repetition without exhaustion. A psalm recited daily remains alive because it opens rather than closes meaning. Physical laws function similarly. They are repeated across experiments, industries, and generations without losing relevance. Each application reveals something slightly new.

The metaphor should not be pushed too far. Physics is not prayer. Equations do not listen. And yet the posture they invite—careful attention to what holds across difference—aligns closely with contemplative habits. To accept a simple law is to accept constraint. One submits personal expectation to a pattern that does not negotiate.

Submission can feel uncomfortable in an age that celebrates complexity as a mark of sophistication. There is a temptation to distrust anything concise, to suspect that brevity hides ignorance. History suggests the opposite. Brevity, when earned, signals endurance. It means a relation has survived contact with the world.

Still, caution remains essential. Not every simple expression deserves reverence. Pseudoscience often mimics elegance without accountability. Numerological schemes promise sweeping insight without testing. What distinguishes a physical law is not its beauty alone but its willingness to fail publicly. Predictions can be wrong. Measurements can contradict. Revision remains possible.

This brings us back to a quiet motif threading this chapter: compression. Folding a vast cloth into a small space risks creasing. Some details disappear. Others align. The success of physics lies in choosing folds that preserve what matters for action while acknowledging what is lost. Engineers know this instinctively. When designing a bridge, they ignore atomic vibrations and focus on forces that matter at scale. The resulting structure stands because the chosen compression matches the task.

The ethical implication is clear. When we mistake compression for totality, we err. When we treat laws as lenses rather than cages, we prosper. This distinction matters far beyond physics. Economic models compress behaviour. Medical guidelines compress populations into averages. Educational metrics compress learning into scores. In each case, elegance can serve or distort depending on humility.

Returning to physical law keeps the lesson grounded. The short relations that describe motion, energy, or fields do not claim to exhaust reality. They claim reliability within bounds. That modesty is their strength. It allows them to travel across centuries without demanding allegiance beyond evidence.

There is also a temporal dimension worth noticing. Simple laws age well. They remain readable even as contexts change. Newton's descriptions still guide spacecraft navigation, corrected where necessary but not discarded. Kepler's insights still underpin orbital mechanics. Maxwell's unification still powers modern technology. Longevity is not proof of truth in any absolute sense, but it is a mark of resonance between language and world.

This resonance invites a question that lingers rather than resolves. When we encounter a relation so compact that it seems to whisper rather than shout, are we witnessing a property of nature or a preference of the human mind? Do we find elegance because the universe is economical, or because our cognition rewards economy? The question remains open. What matters is that the encounter changes how we move.

A short law, trusted, alters behaviour. A satellite is launched. A bridge is crossed. A prediction is made and verified. The law earns its keep by guiding action without demanding comprehension of every underlying detail. That guidance is a gift that carries responsibility.

As this chapter proceeds, the reader is invited to notice a personal response. When encountering these compressed descriptions, do you feel relief or suspicion? Does simplicity feel like clarity or like loss? The elegance of simple laws does not insist on admiration. It invites discernment. The universe seems willing to speak briefly. The task is to listen without assuming the conversation is finished.

Constants that shape reality

On 16 November 2018, in Versailles, delegates to the General Conference on Weights and Measures (CGPM) voted to redefine the kilogram, the ampere, the kelvin, and the mole by fixing exact numerical values for fundamental constants. The organisation, date, and place are easily verified. The concrete outcome was practical and philosophical at once: the world's measurement system was anchored not to physical artefacts—no more platinum-iridium cylinder in a vault—but to stable features of nature such as the Planck constant and the speed of light. In effect, civilisation voted to trust the universe's own steadiness.

A human unit became a cosmic promise.

That decision is a good doorway into the subject because it reveals what constants are for. A physical constant is not merely a number in a table. It is a hinge between theory and world. It is what makes repetition possible across laboratories, continents, and centuries. Without constants, measurement would drift, technologies would misbehave, and the laws described earlier would be little more than local folk wisdom. Constants stabilise the script.

Five words belong here. Change needs something to hold.

The most familiar constant is almost too famous to feel strange: the speed of light in a vacuum, usually written as c. Its value—299,792,458 metres per second—is now defined exactly as part of the SI system, which means it is not merely measured but used to define the metre itself. This is a subtle twist. We used to measure light with rulers; now we define rulers with light. The behaviour is inversion: unit definitions are grounded in constancy. The metric is time of flight. The implication is global coherence. A metre in Manchester and a metre in Mumbai are tied to the same cosmic clockwork.

Even if one never uses the number, its stability shapes daily life. GPS systems rely on electromagnetic signals travelling at a constant speed; timing errors become positional errors. The decision to treat *c* as exact is therefore not abstract. It enables navigation, telecommunications, and the synchronisation that underwrites modern infrastructures.

Yet the wonder is deeper than utility. The speed of light is not simply "how fast light goes." It is also the speed at which causality propagates in our best physical theories. It sets a limit on how quickly information can travel. It behaves like a ceiling on the universe's ability to update itself. That such a ceiling exists, and is the same everywhere we have looked, is a kind of metaphysical jolt. Reality has a tempo.

Another constant, less famous in popular conversation but no less consequential, is the gravitational constant, *G*. This number governs how strongly masses attract one another. It is notoriously difficult to measure precisely compared with many other constants, a limitation that has become part of its story. Here is a useful correction to naïve reverence: "constant" does not always mean "known with great certainty." It means "treated as fixed in the laws." Our knowledge of its value can be comparatively fuzzy, and still the gravitational attraction it describes remains consistent enough to hold planets in orbit.

This distinction matters. It prevents the concept of constancy from becoming magical. Constants are fixed features of theories that match nature with remarkable reliability. But human measurement has limits. Uncertainty bars appear alongside digits. Precision is earned, not assumed.

A named, checkable example makes the point vivid. In 1798, in London, Henry Cavendish conducted his torsion balance experiment to measure Earth's density, effectively performing what became known as the first measurement of *G*. The date and place are verifiable in the history of physics. The outcome was numerical: a method for quantifying gravitational attraction in the laboratory rather than inferring it from celestial motion. Cavendish's apparatus—lead spheres, a suspended rod, a delicate twist of wire—took an invisible force and rendered it measurable. The behaviour was patient measurement. The metric was tiny angular displacement. The implication was profound: the same gravity that rules planets could be weighed on a workbench.

This is one of the recurring miracles of constants. They collapse scales. The same number links the intimate to the cosmic. It connects falling apples to orbital mechanics. It binds the weight of a body to the curvature of spacetime. It makes the universe continuous rather than patchy.

In the quantum realm, the Planck constant, *h*, plays a similar role. It sets the scale at which quantum effects become unavoidable. In ordinary life, objects seem to have definite positions and velocities. At very small scales, those assumptions break. *h* quantifies that break. When the SI system fixed the value of the Planck constant in 2018, it quietly declared that quantum structure is not a fringe curiosity but part of the bedrock on which measurement itself rests.

Here abstraction meets behaviour. A laboratory calibrates a mass standard using electrical measurements tied to quantum effects. The metric is voltage and frequency relationships in devices like Josephson junctions and the quantum Hall effect. The implication is operational solidity: even mass, once linked to a metal cylinder, can be grounded in stable, repeatable quantum phenomena. This is not spirituality dressed as science. It is engineering with metaphysical consequences.

The constancy of these numbers raises a question that the chapter's title insists upon: do constants "shape reality", or do they simply describe it? The phrase is deliberately ambiguous.

A constant in an equation influences what solutions are possible. Change the constant and the world described by the equation changes. That seems to suggest that constants *do* shape reality. Yet from another angle, the constant is our label for a pre-existing feature: the universe behaves as it does, and we are discovering the numbers that capture that behaviour. The constant shapes our description, and our description shapes our interventions. Which direction is fundamental remains contested.

Cosmology intensifies this tension through the notion of fine-tuning. Many constants appear to have values that permit stable atoms, long-lived stars, and complex chemistry. Slight changes, in some models, would yield a radically different universe. Some thinkers see this as evidence of design. Others interpret it through selection effects: we observe a universe compatible with observers because no other kind could be observed. Still others propose multiverse scenarios where many sets of constants exist, and we inhabit one among them.

It is crucial to keep the epistemic boundary clear. These are scenarios, not settled facts. The data constrain some aspects of cosmological models; they do not conclusively answer why constants take their values. This is the limitation that honest reverence must include. Wonder should not outpace evidence. Equally, evidence should not be used to smother wonder.

A second named, checkable example illustrates how constants have been tested, refined, and sometimes suspected of changing. In 1960, the physicist Robert Dicke at Princeton argued for rigorous scrutiny of whether fundamental constants might vary over cosmic time, stimulating lines of inquiry that continue through astrophysical observations and laboratory experiments. The date and institutional context are verifiable in the history of cosmology. The outcome was methodological vigilance: constants are treated as constant because evidence supports it, and they are monitored because science does not sanctify assumptions.

This vigilance is itself a kind of spiritual discipline, in the best sense: a refusal to turn useful ideas into idols. The scientific community's willingness to test whether constants might vary is a model for how to hold claims about the universe with both commitment and openness.

The everyday world offers a quieter route into the same theme. Consider the boiling point of water. It is not a fundamental constant in the same way as c or h, because it depends on pressure and impurities, yet it functions as a familiar anchor. At sea level, water boils around 100°C. At higher altitudes, it boils at lower temperatures. The behaviour is visible in cooking and weather. The metric is temperature. The implication is a lesson in conditionality: even "fixed" facts in daily life carry hidden parameters.

This example is not a diversion. It teaches discernment. Some constants are definitional within our measurement system. Some are empirical features of nature. Some are context-dependent regularities masquerading as constants until conditions change. The mind tends to blend these categories. Healing our relationship with number, as explored in the previous chapter, includes learning such distinctions. Reverence without discrimination becomes superstition. Discrimination without reverence becomes cynicism. The middle path is attentive precision.

The motif threading this subchapter is the anchor. An anchor holds a vessel steady while the sea changes. Constants are anchors in the ocean of phenomena. They do not stop motion. They allow navigation. Without them, one could not measure drift because there would be no reference point.

The ethical implications of anchors are often overlooked. When a society chooses which numbers to treat as foundational—whether in physics, economics, education, or health—it chooses what will be stable and what will be negotiable. The 2018 SI redefinition is therefore a parable about governance. We decided to anchor measurement in nature rather than in a human-made object subject to wear, theft, or accident. We chose stability over convenience, universality over locality. That is a moral choice expressed as metrology.

At the same time, constants remind us of limits. They set ceilings and floors. They constrain what is possible. One cannot build a signalling system that outruns light. One cannot ignore quantum structure when designing nanoscale devices. These constraints are not oppressions. They are the rules of the game. Accepting them is not submission to tyranny; it is agreement with reality.

The forward test is simple but not trivial. The next time you encounter a "constant" in your own life—a fixed number you treat as immovable—ask what kind of constant it is. Is it a true anchor, like a physical limit? Is it a definitional choice, like a unit standard? Or is it a conditional regularity that changes when context shifts? The answer will tell you whether you are reading nature's steadiness, or mistaking habit for law.

Are we reading or writing the code?

On 25 September 2016, in the Basque town of Zumaia, the Laser Interferometer Gravitational-Wave Observatory (LIGO) and its European partner Virgo announced the first direct detection of gravitational waves, a signal known as GW150914 that had been recorded the previous year. The names, dates, and institutional setting are verifiable. The concrete outcome was not merely a headline; it was a vindication of a prediction made a century earlier from a set of equations in general relativity. A ripple in spacetime, inferred by mathematics long before it was seen, registered as a tiny distortion—smaller than a proton's width—across a four-kilometre instrument.

The universe answered.
But whose language was it?

That question is the hinge of this subchapter. When an equation written by a human mind predicts a phenomenon no human has ever experienced directly, and then the phenomenon arrives exactly as the equation described, it tempts us toward a dramatic conclusion: perhaps we are *reading* a pre-existing script. Perhaps the world is written in mathematical code, and science is decipherment.

Yet an equally plausible thought haunts the same story. The prediction was made in a language we invented: symbols, definitions, coordinate systems, and modelling assumptions. Detectors were designed around those assumptions. Data were filtered through templates generated by the theory. The "signal" became meaningful because we knew what to look for. Perhaps, then, we are also *writing* the code—imposing structure on noise, crafting a grammar that fits the world because we prune what does not fit.

Five words belong here. Both can be partly true.

This is not a timid compromise; it is a serious claim about how knowledge works. Science is neither pure discovery nor pure invention. It is a partnership between the world's regularities and the mind's capacity to formalise them. The question "reading or writing?" matters because it governs our spiritual posture. If we are only reading, the universe is primary and we are receivers. If we are only writing, the universe becomes clay and we are sovereign. Neither posture fits the evidence or the ethics of science.

Engineering offers the most relentless test of this debate because it refuses comforting ambiguity. A bridge either stands or fails. A spacecraft either arrives or misses by millions of kilometres. The behaviour is construction. The metric is tolerance: millimetres, microseconds, stress limits. The decision is design based on mathematical models. The implication is accountability. If mathematics were merely a human projection, engineering would be a theatre of continuous catastrophe.

Take an example that is both famous and checkable: the navigation of interplanetary spacecraft. The successful arrival of NASA's Mars Pathfinder in July 1997 depended on precise modelling of trajectories, atmospheric entry, and timing. The organisations, month, and mission details are well documented. The outcome—landing on Mars—was not achieved by intuition or metaphor. It required predictive mathematics that held across a distance of hundreds of millions of kilometres. Here discovery seems undeniable. The world behaved according to relations we did not negotiate.

And yet the same example reveals invention at work. Engineers did not "read" a pre-written trajectory. They chose coordinate frames, approximations, numerical methods, and simplifying assumptions. They decided which forces mattered and which could be neglected. They defined error budgets and built correction manoeuvres. The final success was not proof that our equations are God's handwriting; it was proof that our chosen representation was adequate for the task.

Name the claim. Scientific knowledge is model-dependent. Explain it. A model is a simplified representation that captures relevant structure while ignoring other details. The implication is twofold. First, models can be astonishingly effective. Second, effectiveness does not guarantee metaphysical completeness.

This is why physics has a peculiar texture: it feels like revelation and craft at once. The revelations are real—constants that do not budge, symmetries that persist, relations that hold from laboratories to galaxies. The craft is real too—choosing how to measure, how to formalise, how to compress. The scientist is both reader and author, and the text is written partly in nature and partly in method.

Prediction is where the question becomes sharpest. A prediction is not the same as a post hoc fit. It risks being wrong publicly. When it succeeds, it feels like discovery. When it fails, it reminds us that we are inventing frameworks, not extracting certainty.

The history of Neptune is an instructive case. In 1846, in Berlin, Johann Gottfried Galle observed Neptune following calculations by Urbain Le Verrier, who had inferred the planet's existence from irregularities in Uranus's orbit. The dates, names, and places are checkable. The concrete outcome was astronomical: a planet found because a mathematical model demanded it. This is one of the most persuasive arguments for "reading" the code. Mathematics seemed to reveal an object before the eye could.

Yet even here the story is not pure. The calculation depended on Newtonian gravity, a theory later revised by Einstein. The planet was real, but the framework that led to it was an approximation. Discovery occurred through an invented lens. The lens worked well enough to point the telescope. That is the proper moral.

A credible limitation must now be faced. Not all predictive successes imply that mathematics is embedded in nature as an independent code. Sometimes prediction succeeds because we choose the framework that makes it succeed, and ignore alternatives. This is not conspiracy; it is selection. Scientists prefer theories that are simple, coherent, and predictive. Those aesthetic and pragmatic preferences shape what survives. The result can look like the universe is "written mathematically" when in fact we have filtered the available descriptions through human criteria.

Counter-examples also exist where models predict brilliantly and still mislead. Classical mechanics predicts most everyday motion with uncanny accuracy while remaining conceptually incomplete at small scales and high speeds. Newton's laws were not "wrong" in ordinary contexts; they were limited. Likewise, economic models can predict trends under certain conditions while failing catastrophically in crises because human behaviour shifts. Prediction is not a guarantee of ontological truth. It is evidence of fit within a domain.

Reconciliation requires a distinction between *structure* and *notation*. The structure appears to be real. The notation is ours. Symmetry in physical laws seems not to depend on our preferences. Conservation of energy, for example, manifests across experiments in ways that are stubbornly consistent. Yet the way we express that conservation—through calculus, diagrams, or computational models—is human-made. A different intelligence might express the same structure differently, but it would still need to respect the structure to build anything that works.

This view aligns with a sober philosophy of science. In 1960, the physicist Eugene Wigner published his essay "The Unreasonable Effectiveness of Mathematics in the Natural Sciences." The year, author, and title are verifiable. The concrete outcome was conceptual: he gave voice to a perplexity many scientists feel but rarely articulate. Mathematics, a human creation, seems uncannily suited to describing the world. Wigner did not settle the question. He sharpened it.

The phrase "unreasonable effectiveness" has been interpreted in contradictory ways. Some treat it as evidence that reality is fundamentally mathematical. Others treat it as evidence that mathematics is flexible enough to model almost anything, and we remember the successes more vividly than the failures. Both interpretations carry weight, and neither dissolves the other.

This is where the spiritual dimension enters without requiring theology. If the world is structured in ways that can be grasped, that graspability demands humility rather than conquest. It suggests that human intelligence is not alien to the cosmos. It is, at minimum, compatible with it. That compatibility can be framed as evolutionary accident, cosmic hospitality, or something deeper. The book does not force a single metaphysical reading. It insists, however, that the compatibility is not trivial.

The motif threading this subchapter is the key. A key is shaped by the lock it opens. But the key is also made by a craftsman. Mathematics functions like a key in both senses. It fits the world's regularities. It is also forged—through definitions, axioms, and methods—by human minds. When the key turns cleanly, we are tempted to imagine the lock was designed for it. When it jams, we learn the limits of our craft.

A modern twist complicates the debate further: machine learning. These systems often generate accurate predictions without transparent equations. They "read" patterns in data through optimisation, producing models that work yet resist human explanation. Here we encounter a third category: effective prediction without intelligible code, at least to us. The behaviour is training on data. The metric is predictive accuracy. The implication is unsettling: perhaps the universe can be predicted in ways that do not map neatly onto the elegant, compact laws we prize.

This does not refute the existence of underlying structure. It challenges our assumption that structure must always be expressible in the kinds of shorthand humans find beautiful. The world may have order that exceeds our preferred grammars. If so, humility deepens. We remain readers, but not necessarily fluent.

The forward test is therefore not "Do you believe the universe is code?" but a more discriminating question. When a model predicts successfully, do you treat that success as permission to stop asking what the model ignores? Or do you treat it as a provisional key—useful, limited, and worth refining? Your answer determines whether numbers become instruments of wisdom or instruments of domination.

Humility before the unknown

In December 1900, in Berlin, the physicist Max Planck presented a paper to the German Physical Society proposing that energy is emitted in discrete packets. The date, place, and institutional context are checkable in the history of physics. The concrete outcome was not immediate acclaim but a crack in a confident worldview. Classical physics, which had seemed capable of explaining heat, light, and motion with ever-increasing precision, could not account for blackbody radiation without producing nonsense. Planck introduced a constant—later called *h*—as a mathematical device. Within a generation, it would become a cornerstone of quantum theory.

He was trying to fix an equation.
He altered reality's image.

This is the kind of story science tells at its best: not the tale of smooth accumulation, but of surprise, revision, and reluctant humility. A theory works—beautifully, usefully—until it does not. The failure is not a minor inconvenience. It is a demand. It forces scientists to admit that their best language has reached its edge.

Five words belong here. Knowledge grows by being wrong.

That sentence is uncomfortable, and it should be. Many people approach science as if it were a cathedral of certainty: one stone laid atop another, the structure rising steadily towards truth. The lived practice is more like navigation through fog with instruments that occasionally lie. The miracle is not that errors occur; it is that the process can identify and correct them.

To see why humility is not an optional virtue in science, notice what counts as success. A successful theory does not merely describe what is already known; it risks itself against what is unknown. It makes predictions. It commits. When a prediction fails, the failure is not embarrassment. It is information. The behaviour is disciplined testing. The metric is deviation between expectation and observation. The implication is revision.

Planck's move is instructive because it shows that even a minor-seeming adjustment can have vast consequences. He did not set out to create "quantum mechanics." He sought a way to match theory to measurement. The measurements—meticulous experimental work by people such as Heinrich Rubens and Ferdinand Kurlbaum in Berlin—forced the issue. The constant appeared as a patch. It became a portal.

The spiritual lesson here is not that science is fragile. It is that human knowledge is bounded. When the bounds are acknowledged, rigour strengthens. When the bounds are denied, confidence turns brittle.

Another correction story reveals this brittleness in motion. In 1859, in London, Urbain Le Verrier—already celebrated for predicting Neptune—announced that Mercury's orbit showed an unexplained advance of its perihelion. The date and place of the report are checkable in the scientific record. Newtonian gravity could not fully account for the observed motion. For decades, astronomers proposed fixes: an unseen planet (Vulcan), dust, tweaks to gravitational law. The concrete outcome was persistent discrepancy. The universe refused to conform.

In 1915, in Berlin, Albert Einstein published general relativity, and the discrepancy became a triumph of correction. The mysterious extra motion of Mercury's perihelion emerged naturally from a new account of gravity. The behaviour of scientists across this arc is revealing. They did not abandon

Newton at the first anomaly. They tested, recalculated, proposed alternatives, and waited. They treated the mismatch as a problem worth honouring, not a nuisance to be ignored.

Name the claim. Anomalies are sacred data. Explain it. An anomaly is an observation that resists a prevailing framework. It signals either measurement error or a limitation in the theory. The implication is methodological humility: instead of forcing nature to fit, science lets nature interrogate its assumptions.

Humility, however, is often misunderstood as mere tentativeness. In science it is something sharper: a willingness to commit strongly to the best available model while remaining ready to revise it when evidence demands. That is not indecision. It is disciplined openness.

The twentieth century offers further surprises that underline the point. In 1928, in Cambridge, the physicist Paul Dirac derived an equation combining quantum mechanics and special relativity. The year and place are verifiable. The concrete outcome was startling. The mathematics implied the existence of particles identical to electrons but with opposite charge—antimatter. This was not invented as a metaphysical flourish. It was an unwelcome implication of an elegant framework.

In 1932, in Pasadena, Carl Anderson observed the positron in cosmic-ray tracks, confirming the prediction. Again, we see the peculiar humility of science: it allows mathematics, when constrained by evidence, to lead beyond common sense. Yet it does not declare victory until observation answers. The tension between boldness and restraint is the virtue.

These stories can be misused. Some readers conclude that because theories have been corrected, nothing can be trusted. That is a lazy scepticism, and it mistakes revision for collapse. The better inference is that scientific knowledge is layered.

Older theories remain effective within their domains even when deeper frameworks exist. Newton is not banished; he is bounded. Classical thermodynamics remains reliable even as statistical mechanics explains its foundations. The corrections do not erase; they contextualise.

A credible limitation must be stated. Scientific revision is not always clean or noble. Human incentives distort it. Careers are built on paradigms. Funding prefers fashionable questions. Institutions sometimes resist disruptive ideas. Thomas Kuhn's *The Structure of Scientific Revolutions* (1962) famously analysed how paradigms shape what scientists notice and accept. The book and year are checkable. The concrete outcome was conceptual: a reminder that science is practised by humans, not angels.

This limitation does not discredit science. It clarifies what humility must include: not only humility before nature, but humility about our social machinery. A community can become attached to its models. Peer review can become conservative. Consensus can be useful and still be wrong.

Reconciliation lies in the structure of the scientific method itself. The method is designed to allow nature to veto human preference. Experiments can be repeated. Data can be re-analysed. Predictions can be tested independently. This does not guarantee virtue, but it creates pathways for correction even when psychology and politics resist.

Modern cosmology provides an example of surprise that is both recent and stabilised in the record. In 1998, two independent teams—the Supernova Cosmology Project and the High-Z Supernova Search Team—reported evidence that the expansion of the universe is accelerating, based on observations of distant Type Ia supernovae. The year and the team names are verifiable. The concrete outcome was a shock: gravity, assumed to slow expansion, seemed outweighed by a repulsive component later associated with dark energy or a cosmological constant.

This surprise forced a re-reading of the universe's story. It also revealed how much remains unknown. The majority of the cosmos, in standard models, consists of dark matter and dark energy—entities inferred from gravitational effects rather than directly observed in laboratory form. The numbers add up. The substances remain mysterious. Humility here is not pious decoration; it is the honest state of the field.

At this point, it is tempting to lapse into a romantic hymn to mystery. That would be a mistake. Mystery in science is not a licence for vague speculation. It is a map of unanswered questions shaped by constraints. Dark matter is not a poetic placeholder; it is a name for measurable gravitational behaviour that does not match visible matter. The behaviour is galactic rotation and lensing. The metric is velocity curves and light bending. The implication is disciplined ignorance: we know something is happening, and we know we do not yet have the right description of what.

A quieter form of humility appears in mundane laboratory life. Constants are refined. Error bars are reported. Results are replicated. Negative results are wrestled with. The discipline is not glamorous, but it is central. A scientist who reports an uncertainty is not confessing weakness; they are describing the boundary of what their measurement supports. This boundary is part of the truth.

The spiritual dimension emerges when one considers how this discipline trains the self. To live with uncertainty without fleeing into fantasy or rigid certainty is a mature human skill. Science, practiced honestly, cultivates it. It teaches patience. It teaches restraint in claiming. It teaches reverence for what resists capture.

The motif threading this subchapter is the horizon. A horizon is not a wall. It is a boundary of visibility. As one moves, it moves. Scientific knowledge behaves similarly. Each advance reveals further unknowns. The horizon recedes. The temptation is either triumphalism ("we are nearing the end") or despair ("we know nothing"). Humility refuses both. It accepts progress without pretending to finality.

This is especially urgent now because scientific authority is both needed and contested. Public debates about climate, health, and technology often demand certainty where none exists. Models provide probabilities, not prophecies. Uncertainty is interpreted as weakness rather than honesty. This social pressure incentivises overstatement, and overstatement eventually corrodes trust when revisions occur. Humility is therefore not merely a personal virtue. It is a public responsibility.

To combine intellectual rigour with spiritual humility is to adopt a particular posture towards knowledge. Rigour means checking claims, quantifying uncertainty, demanding evidence, and refining models. Humility means acknowledging limits, resisting idolisation of current frameworks, and remaining open to surprise. The two belong together. Rigour without humility becomes tyranny of intellect. Humility without rigour becomes a swamp.

The forward test is simple to phrase and difficult to enact. When you encounter a confident scientific claim—your own or someone else's—ask what would count as a meaningful falsification. What observation, what measurement, what pattern would force revision? If no answer can be named, the claim may be ideology wearing a lab coat. If an answer can be named, you are in the presence of genuine knowledge: committed, vulnerable, and alive.

Chapter 9
Counting as Contemplation

At dawn in Kyoto, the wooden floor of a Zen temple registers a soft rhythm: step, pause, step. Each movement is counted, not aloud, not for speed or score, but as a way of arriving. The practice is centuries old. Elsewhere, fingers move across beads worn smooth by generations of prayer; breath aligns with syllables repeated a fixed number of times; pilgrims measure distance not in miles but in steps. Nothing is being optimised. Nothing is being tracked for improvement. Counting here serves a different master. It steadies the mind.

This is the paradox at the heart of contemplative counting. Numbers, so often experienced as instruments of pressure, become instruments of release. The same human faculty that tallies profit, calories, followers, and deadlines can also cradle attention, soften urgency, and make room for presence. The shift is not in the digits themselves but in the posture with which they are held.

Modern readers may feel an immediate resistance. Counting has become noisy. Our days hum with metrics: steps walked, emails sent, hours slept, tasks completed. The phone glows with dashboards. The watch vibrates with reminders. In such an environment, the idea that counting could quiet the mind sounds implausible. Yet the practices explored in this chapter persist precisely because they answer a need that has not changed: the need to anchor awareness in the body and the moment.

Historically, counted repetition appears wherever human beings have sought stability amid flux. The rosary in Western Christianity, standardised in its familiar form by the late Middle Ages, structures prayer through sets of ten, punctuated by larger markers. In Hindu and Buddhist traditions, malas of 108 beads guide mantra recitation, the number chosen for symbolic and cosmological reasons that need not be rehearsed here. Islamic prayer beads, often grouped in ninety-nine, echo divine names. Walking meditations in monastic settings use counted steps to prevent the mind from scattering. In each case, number acts as a container.

The behaviour is simple: repeat a word, a breath, a step, a bead. The metric is fixed. The implication is subtle. Attention no longer has to decide what comes next. Choice recedes. Presence grows.

This chapter does not romanticise these traditions or suggest that ancient practices are automatically wiser than modern ones. It asks a narrower, more practical question. What happens when number is used not to evaluate but to accompany experience? What changes when counting is detached from performance and reattached to presence?

The question matters because many contemporary efforts at self-care unknowingly reproduce the very anxieties they aim to soothe. Gratitude becomes a checklist. Meditation becomes streaks and minutes logged. Exercise becomes rings closed. The intention may be wholesome. The effect can be corrosive. Counting clarifies until it constricts.

Here, a quieter approach is proposed. Small, humane tallies—three moments of gratitude noted at day's end, five acts of kindness observed rather than broadcast, ten slow breaths before a difficult conversation—are not presented as techniques to maximise happiness. They are gestures. They make attention visible without demanding perfection. Their scale is deliberate. They fit into ordinary days without ceremony or equipment.

Notice the difference. A daily gratitude journal that demands novelty and eloquence soon becomes a burden. A simple habit of noticing three things that did not need to happen but did—a warm cup, a kind glance, a task completed—can soften the edges of perception. The behaviour is noticing. The metric is modest. The implication is relief.

This modesty is not accidental. It resists the modern compulsion to refine endlessly. Much harm arises not from measurement itself but from refusing to stop measuring. The last digit becomes an obsession. Sleep is not enough unless it is optimal. Work is not sufficient unless it is quantified. Health is not felt unless it is graphed. Precision, once a servant, becomes a tyrant.

One of the skills cultivated in contemplative counting is learning when to let go of the final refinement. Many spiritual traditions recognise this moment intuitively. Prayer ends not when the count is perfect but when the time is right. A walk concludes when the body signals completion, not when a target is met. Acceptance of the approximate is not sloppiness. It is trust.

This chapter invites readers to observe, gently and without judgement, their own urge to quantify. When does counting bring clarity? When does it add weight? The invitation is observational rather than prescriptive. Watch the hand reach for the calculator of the self. Notice the impulse to compare steps, hours, outputs. Ask what is being sought: reassurance, control, worth.

Such noticing is itself a form of contemplation. It turns number from an external authority into an object of awareness. The mind learns to see its habits. Freedom begins there.

There is also a social dimension. Counting together—steps taken in silence, breaths shared in a group, beads passed through communal hands—creates synchrony. Neuroscience has its explanations for this, but the lived experience is older and simpler. Shared rhythm binds without speech. Number becomes a bridge rather than a scorecard.

The motif threading this chapter is the pulse. Breath has one. Steps have one. Days have one. Counting, when aligned with pulse rather than imposed upon it, amplifies life instead of fragmenting it. The reader is not asked to adopt new identities or beliefs. Only to experiment with posture: to see what changes when number accompanies rather than commands.

As the chapter unfolds, ancient practices and contemporary habits will be placed side by side, not to judge but to illuminate. The aim is not to escape numbers, an impossible task in any functioning society, but to rehabilitate them. To remember that counting can be an act of care.

The unresolved tension is this. Numbers will continue to multiply around us. Metrics will not retreat. The decision lies elsewhere. Will counting be another arena of self-surveillance, or can it become a quiet form of attention? The pages that follow do not answer this once and for all. They offer ways of noticing. The rest is lived.

Breath, beads, and steps

In August 1982, at Plum Village in southwest France, Thích Nhất Hạnh introduced a form of walking meditation to Western retreatants that was almost disarmingly plain. Each step was taken in time with the breath. Each breath was silently counted. No posture correction. No visualisation. The outcome, recorded in retreat notes and later teachings, was not mystical transport but something more modest and harder to fake: participants reported steadier attention, fewer intrusive thoughts, and a surprising ease in sustaining silence for long periods. Nothing new was invented. An old practice had simply been named, slowed, and shared.

Count the breath.
That is all.

The scene matters because it cuts against a common assumption. Repetition, especially counted repetition, is often dismissed as mechanical, even mind-numbing. The word "rote" lingers nearby. Yet across cultures and centuries, counted practices appear wherever people have sought to quiet the mind without suppressing it. Rosaries, malas, prayer ropes, knotted cords, paced footsteps, and measured bows all do the same unglamorous work. They give attention a rail to run on.

Consider the Catholic rosary as it emerged in its recognisable form in Western Europe by the fifteenth century. Its structure is precise: decades of repeated prayer marked by larger beads, with a total count that can be verified by touch alone. The behaviour is physical. Fingers move. Lips form familiar words. The metric is fixed. The implication is freedom from choice. One does not decide what comes next. The prayer arrives already shaped.

This matters because decision-making is cognitively expensive. Modern psychology would later give names to this cost, but the practitioners did not need the terminology. They learned empirically that reducing choice reduces agitation. A rosary prayed poorly still counts. A rosary prayed distractedly still holds the hands. The number does not judge the quality of attention. It shelters it.

A similar logic governs the use of malas in Hindu and Buddhist traditions, commonly composed of 108 beads. The number itself carries layers of symbolic meaning, but the practical effect lies elsewhere. A practitioner recites a mantra once per bead, moving forward until the circuit completes. The visible behaviour is repetition. The metric is tactile. The implication is endurance without strain. Attention does not have to expand heroically. It only has to stay.

There is a temptation to romanticise such objects, to treat them as exotic artefacts rather than tools. That temptation obscures their most important feature: they outsource counting. The mind does not need to remember where it is. The hand remembers. When attention wanders, the body keeps the place.

Walking meditation extends this logic into space. In Zen monasteries, kinhin—slow walking between periods of seated practice—often uses counted steps coordinated with breathing. Two steps on the in-breath. Three on the out-breath. The exact ratio varies by school, but the principle holds. The body becomes a metronome. Movement becomes measure.

Here, number stabilises not thought but motion. The feet learn the pace. The breath settles. The mind, deprived of novelty, begins to loosen its grip. This is not transcendence. It is regulation.

Modern neuroscience has begun to describe what long practice already knew. Rhythmic, predictable activity reduces variability in attention networks and dampens stress responses. Studies of paced breathing show measurable effects on heart rate variability, a marker associated with emotional regulation. When breath is counted—say, four counts in, six counts out—respiration slows, carbon dioxide tolerance increases, and the nervous system shifts. The behaviour is counting. The metric is time. The implication is physiological change.

Yet it would be a mistake to claim that counted practices always soothe. There is a credible counter-example. For some individuals, especially those prone to compulsive tendencies, counting can become a source of tension rather than release. The metric tightens. Missing a count induces anxiety. The practice hardens.

This limitation matters. It clarifies what makes contemplative counting work. The number must serve attention, not dominate it. In traditions that have endured, this distinction is explicit. A Zen teacher will instruct a student to return gently to the count when it is lost, not to start again in self-reproach. Catholic spiritual directors have long cautioned against scrupulosity, the obsessive fixation on correctness. The number is a guide, not a judge.

Reconciled properly, this limitation sharpens the point. Counting softens the mind only when it is held lightly. The structure is firm. The attitude is forgiving.

The quiet motif threading these practices is the hand. Fingers on beads. Soles on earth. Palm rising and falling with breath. The hand knows how to repeat without comment. It does not evaluate progress. It simply continues.

This embodied knowledge stands in contrast to much modern engagement with numbers. Fitness trackers count steps, but they also compare them. Meditation apps count minutes, but they also rank streaks. The same numerical act—counting breaths, steps, repetitions—acquires a different moral weight. The behaviour is identical. The implication diverges.

In a walking meditation, ten steps are ten steps. No more. No less. In a tracked walk, ten steps are negligible. They barely register. Meaning migrates from experience to display.

The distinction is subtle but decisive. Contemplative counting keeps the number close to the body. Instrumental counting exports it to a system.

This helps explain why so many people report a sense of relief when returning to untracked practices. Counting beads without logging them. Walking without recording distance. Breathing without timing an interval. The number is still present. It simply does not accumulate into identity.

Ancient sources often describe this relief indirectly. In the Philokalia, a collection of Eastern Christian texts compiled in the eighteenth century but drawing on much earlier material, the Jesus Prayer is sometimes coordinated with breath and repeated continuously. The goal is not to reach a certain count but to let the prayer "descend into the heart". The metric disappears into habit. Counting gives way to rhythm.

This transition is instructive. Counting is not the destination. It is the entry point. It steadies attention until attention can stand on its own.

The same pattern appears in secular adaptations. A clinician teaching paced breathing to manage anxiety may begin with explicit counts. Inhale for four. Exhale for six. Over time, the patient internalises the rhythm. The numbers fall away. What remains is ease.

The implication for modern readers is not that they should adopt any particular tradition. It is that counted repetition works best when it is temporary, humble, and embodied. A small number repeated well outperforms a large number chased anxiously.

Three breaths taken with care can interrupt a spiral of reactivity. Ten steps walked slowly can reset a meeting gone stale. A single circuit of beads can mark a boundary between tasks. These are not achievements. They are pauses.

The unresolved tension lies here. Our culture prizes accumulation. More steps. More minutes. More days in a row. Contemplative counting asks a quieter question. How few repetitions are enough to return? How little structure is required to soften the grip?

The answer varies. It cannot be standardised without betraying the point. The practices endure because they resist optimisation. They invite repetition without demand. Count the breath. Touch the bead. Take the step. Then notice what happens when the number does not ask for anything in return.

Small daily tallies

In November 2003, psychologists Robert Emmons and Michael McCullough published a paper from the University of California, Davis, reporting on a simple intervention. Participants were asked to record a small number of things for which they felt grateful each week. The outcome was unexpectedly concrete. Compared with control groups, those keeping the gratitude tally reported better mood, fewer physical complaints, and greater optimism about the coming week. The place was a laboratory. The method was ordinary. The result travelled.

Count three things.
Stop.

The power of the finding lay not in novelty but in scale. Three items. Once a week. No narrative expansion required. No deep excavation of feeling. The behaviour was writing. The metric was small. The implication was sustainable attention.

Small tallies work because they fit the grain of daily life. Large numerical projects provoke resistance. They demand time, memory, and motivation. Modest counts slip through. They can be completed on a bus, at a kitchen counter, or in the margin of a notebook already in use. They ask for presence, not transformation.

This is not accidental. Across traditions, practices that endure tend to be parsimonious. Jewish ethical texts often recommend a small, fixed number of daily acts to anchor intention. In the Pirkei Avot, a foundational rabbinic collection compiled around the third century in Roman Palestine, ethical instruction is frequently framed in countable units. One deed leads to another. One kindness outweighs many words. The numbers are low. The memory load is light.

The same principle appears in modern secular contexts. Behavioural economists have long observed that habits with low activation energy persist. A tally of five small acts of kindness in a day is measurable without becoming theatrical. Holding a door. Answering an email promptly. Thanking a colleague by name. The behaviour is visible. The metric is finite. The implication is attentiveness to social texture.

Notice what is not required. No ranking of kindnesses. No evaluation of impact. The tally does not ask whether the act was efficient or optimised. It asks whether it happened.

This restraint matters because numbers are easily recruited into self-surveillance. A daily tally can turn punitive. Miss the count and the day is judged a failure. This is the credible counter-example. When the number hardens into a standard, the practice curdles.

The reconciliation lies in how the tally is framed. In Emmons and McCullough's study, participants were not instructed to reach for profound gratitudes. Mundane items counted. A decent meal. A conversation that did not sting. The number invited noticing, not performance.

Here the quiet motif emerges: the notebook. Not the dashboard. Not the app. A page with space for three lines. The notebook does not calculate trends. It receives entries.

Small tallies also function as filters. They direct attention without exhausting it. Consider the practice of noting three moments of beauty encountered in a day. The behaviour is noticing. The metric is capped. The implication is selectivity. Because only three will fit, attention sharpens. The mind begins to choose.

This selective pressure has consequences. Studies on attentional bias suggest that what is repeatedly noticed becomes easier to notice again. A capped tally trains perception. It does not attempt to capture the day. It samples it.

Contrast this with exhaustive tracking. Logging every calorie, every step, every minute of focus produces a different cognitive climate. The metric expands. Attention fragments. The tally ceases to be contemplative.

There is a historical echo here. In the Rule of St Benedict, written in sixth-century Italy, the daily rhythm of monastic life is structured by repeated, limited counts. Fixed hours of prayer. A measured distribution of psalms across the week. Benedict's concern is explicit. Excessive strain breaks the practice. Moderation preserves it.

Small numbers endure because they forgive omission. Miss one gratitude and the practice remains intact. Skip a kindness and the tally resets tomorrow without drama. The number does not accumulate into debt.

This feature distinguishes contemplative tallies from targets. Targets imply consequence. Tallies imply orientation.

A contemporary example illustrates the distinction. In some schools, pupils are asked to record a fixed number of reading minutes per day. The behaviour is reading. The metric is time. The implication, however, often becomes compliance rather than engagement. When the minute count replaces the experience of reading, children learn to watch the clock.

Contrast this with a small tally of pages that lingered in memory. The behaviour shifts. The metric remains small. The implication is meaning.

The difference is subtle. Both involve numbers. One colonises the activity. The other frames it.

This framing is why modest numerical rituals adapt easily to crowded lives. A professional moving between meetings can still note three gratitudes while washing hands. A parent can register five kind acts without extending the day. The tally does not demand a special setting.

Yet the risk remains. Numbers attract comparison. One person's three gratitudes feel thin beside another's eloquence. This social drift can undo the practice.

Here again, restraint rescues it. Tallies work best when they remain private or lightly shared. The notebook stays closed. The count is for attention, not display.

A second named example sharpens the point. In 2014, the UK charity Action for Happiness popularised the "Three Good Things" exercise through public campaigns and workplace programmes. Evaluations conducted with partner organisations reported improved wellbeing scores over short periods. The place varied. The method did not. Three items. End of day. No embellishment.

The counter-example appeared quickly. In teams where the exercise was monitored or reported upward, participation dropped. The behaviour was the same. The implication changed. What had been reflective became performative.

Reconciled properly, the lesson is clear. Small tallies should terminate in the person, not the system. They are tools for alignment, not evidence.

This insight extends beyond gratitude and kindness. Some people keep a tally of moments when they paused before reacting. Others note a fixed number of times they returned to the breath. Others record one question that stayed with them through the day. The content varies. The structure holds.

The hand returns to the motif of the notebook. A single page per day. Enough space for five short marks. The notebook does not ask for consistency across months. It holds today.

The closing tension remains unresolved by design. Numbers clarify. They also constrain. A small tally can sharpen attention or shrink it. The same count can open the day or judge it.

The forward test is simple to state and difficult to answer. When you choose a number for a daily ritual, does it release attention or recruit it into another obligation? The answer is not theoretical. It is felt, quietly, in the moment when the last line is written and the page is closed.

Noticing the urge to quantify

In April 2015, at Apple's headquarters in Cupertino, California, the company released the Apple Watch to the public. Its concrete outcome was immediate and measurable. Within weeks, millions of wrists began lighting up with rings to close, steps to complete, calories to burn. The behaviour was not new. People had walked, slept, and exercised before. What changed was the reflex. Movement now arrived with a number attached, refreshed minute by minute.

The watch did not invent counting. It normalised it. And it made the numbers portable.

This portability matters because it reveals a deeper habit. Humans do not merely use numbers. We reach for them, often without noticing. The urge to quantify precedes deliberation. Before asking whether a count is helpful, the mind often performs it automatically. How many emails answered. How many steps taken. How many likes received. The number arrives first. Meaning follows, if at all.

The first task, then, is not to abandon counting but to observe its arrival. Watching the moment when a number is summoned exposes its purpose. Sometimes it clarifies. Sometimes it burdens.

Consider productivity. A professional opens a task manager at the start of the day. The behaviour is listing. The metric is the number of tasks completed. The implication appears neutral. Yet by midday, the list begins to shape choices. Short tasks are favoured. Complex work is deferred. The count has become a compass.

This is not inherently wrong. In environments where throughput matters, counts protect against drift. Warehouses measure pick rates. Hospitals track waiting times. In these contexts, the metric is paired with a visible decision. Staffing levels adjust. Processes change. Counting serves action.

The trouble begins when the same logic migrates inward. An individual begins to count tasks as a proxy for worth. Ten completed items feel virtuous. Three feel suspect. The behaviour remains checking boxes. The implication shifts to self-evaluation.

This shift is subtle. It is rarely announced. It emerges through repetition.

A second example makes the dynamic plain. In 2009, Facebook introduced the "Like" button at its Menlo Park campus. The feature's stated aim was simple feedback. The outcome was a visible counter attached to social expression. The behaviour became posting. The metric became likes received. The implication quickly expanded into comparison.

Researchers at the University of Michigan, led by Ethan Kross, reported in 2013 that frequent Facebook use correlated with declines in subjective wellbeing over time. The mechanism was not content itself but comparison. The number made the comparison frictionless.

Here, the abstraction is envy. The visible behaviour is refreshing the feed. The metric is the like count. The decision is whether to post again, delete, or withdraw. Counting did not merely report social response. It organised attention around it.

Yet numbers are not villains. They are instruments. The same device that provokes comparison can support care. Step counts help patients recovering from surgery gauge progress. Blood glucose numbers guide diabetics in daily decisions. In these cases, the count clarifies a real constraint.

The difference lies in intention and scope. Does the number answer a question that cannot be answered otherwise? Or does it substitute for a harder judgement?

Health offers a fertile testing ground. Many people track sleep with wearable devices. The behaviour is wearing the watch. The metric is hours slept, sometimes broken into phases. The implication should be rest. Yet for some, the number induces anxiety. A night that felt restorative is judged insufficient by the app. The person wakes tired because the graph says so.

This phenomenon has been documented. Sleep researchers have coined the term "orthosomnia" to describe an unhealthy fixation on achieving perfect sleep metrics, reported in clinical literature around 2017. The abstraction is restfulness. The visible behaviour is checking the sleep score. The decision becomes compensatory napping or worry. The number overrides sensation.

The counter-example is instructive. In clinical settings, sleep diaries often ask patients to record subjective sleep quality alongside duration. The count is paired with feeling. The metric is contextualised. The implication returns to lived experience.

This pairing suggests a principle. Counting clarifies when it is tethered to a purpose outside itself. It burdens when it becomes self-referential.

The quiet motif threading this chapter is the glance. A quick look at a screen. A flicker of a number. The glance seems harmless. It is brief. Yet repeated hundreds of times a day, it trains the mind to seek validation or correction from digits.

Watching the glance is the work. Not suppressing it. Not moralising it. Simply noticing when it happens.

Productivity culture provides another lens. In 1911, Frederick Winslow Taylor published *The Principles of Scientific Management* in the United States, advocating time-and-motion studies to improve industrial efficiency. The outcome was measurable productivity gains in factories. The place was the shop floor. The method was counting.

Taylor's system assumed external observation. Workers were measured by supervisors. Today, the supervisor often sits inside the device. People time their own focus sessions, count their own outputs, and rank their own days. The abstraction is efficiency. The behaviour is self-tracking. The implication can become self-surveillance.

Yet even here, reconciliation is possible. Some writers use word counts to protect creative time. The number sets a boundary. Write for forty-five minutes. Stop. The metric limits intrusion rather than extending it. The count serves the work.

The same tool, different posture.

A credible limitation must be acknowledged. Humans are pattern-seeking. Numbers compress complexity. Without them, many systems would be unmanageable. Public health relies on incidence rates. Climate science depends on long-term averages. To suggest that noticing the urge to quantify should lead to abstinence would be naïve.

The aim is discrimination. When does counting illuminate structure? When does it obscure meaning?

A practical diagnostic emerges. Ask what decision the number enables. If none follows, the count may be ornamental. If the decision is punitive or comparative, the count may be corrosive. If the decision is adaptive, the count earns its place.

This diagnostic applies to self-comparison. A runner checks pace to adjust effort during training. The metric informs action. A runner checks pace to judge worth against others. The metric feeds identity. The behaviour looks identical. The implication diverges.

It also applies to money. Tracking expenses can reveal patterns and support choice. Obsessively refreshing account balances without action can amplify anxiety. The abstraction is security. The behaviour is checking. The metric is the balance. The decision is often absent.

Historical practices remind us that this tension is old. In monastic traditions, monks were cautioned against counting virtues or sins too precisely. Spiritual manuals warned that enumeration could become prideful or scrupulous. The number was to assist examination, not replace it.

This caution resonates with contemporary psychology. Obsessive counting behaviours appear in anxiety disorders, where numbers promise control over uncertainty. The promise is false, but the relief is momentary. The behaviour persists because the number arrives quickly.

Here the motif of the glance returns. The number flashes. Control seems near. The mind tightens.

What interrupts the cycle is not force but curiosity. Why this number? Why now?

This question does not demand an answer every time. It invites a pause. Sometimes the pause reveals a clear purpose. Sometimes it exposes habit.

The chapter's forward movement depends on this exposure. Noticing the urge to quantify prepares the ground for choosing when to count and when to stop. It restores agency.

The unresolved tension remains deliberate. Numbers are indispensable. They are also seductive. The same digit can orient or oppress. The difference lies less in the number itself than in the moment it is summoned.

The forward test is simple and demanding. The next time a number appears unbidden, do you accept it as guidance, or do you ask what question it claims to answer? The answer will not be numerical. It will arrive, quietly, in the quality of attention that follows.

Letting go of the last digit

In January 1986, in Florida, the Space Shuttle *Challenger* broke apart seventy-three seconds after launch. The names, month, year, and place are easily verified. The concrete outcome was catastrophic: seven astronauts died, and the United States suspended the shuttle programme while investigators traced causes through engineering decisions, organisational pressures, and misread risk. One detail remains haunting precisely because it is numerical. Engineers at Morton Thiokol expressed concern about O-ring performance in cold temperatures, but the decision to launch proceeded amid uncertainty and a desire for schedule precision.

Precision did not save them.
It never does, alone.

This is not a cheap analogy for daily life. It is a reminder of a deeper truth about numbers: they can create an illusion of control. The last digit is often where that illusion gathers. We refine. We add decimal places. We narrow uncertainty. We polish the plan until it shines. And then reality does what it always does—it moves.

The aim of this subchapter is not to disparage measurement. It is to name the point at which refinement becomes a substitute for courage. Letting go of the last digit is not laziness. It is discernment: knowing when accuracy improves action and when it merely delays it.

Consider the simplest case: planning a day. A person decides to wake at 6:37, leave at 7:12, answer emails at 8:05, and begin deep work at 9:17. Each timestamp feels precise. The behaviour is scheduling. The metric is minutes. The implication is a promise of smoothness. Yet a child wakes early, the train runs late, an unexpected call arrives. The schedule collapses.

The collapse does not mean planning is futile. It means planning has a proper grain. A plan can hold in quarters of an hour; it rarely holds in minutes. The last digit in the timetable—those final minutes—offers an emotional comfort that the world cannot honour.

This difference between actionable precision and performative precision appears throughout science and engineering. A good engineer knows that measurements come with tolerances. A component can be within spec without being exact. A good experimenter knows that reporting eight decimal places on a value measured with a crude instrument is not impressive; it is dishonest.

Name the claim. Digits imply certainty. Explain it. Each additional digit suggests a reduction in uncertainty. The implication is moral as well as practical: false precision misleads the self and others.

One of the most famous cautions in the philosophy of measurement is that numbers should not claim more than their method supports. This is why scientific papers report error bars and confidence intervals. They are not hedges. They are signals of integrity. They announce where knowledge ends.

Yet outside laboratories, people often treat numbers as absolutes. Diets hinge on single calories. Sleep is judged by a score. Productivity is summarised by a daily count. The more intimate the domain, the more seductive the last digit becomes, because it appears to confer mastery over messy realities.

This seduction has a behavioural signature. The person refreshes the tracker. They check the updated number. They adjust the plan. They check again. The metric becomes a loop. The decision becomes micro-correction. The implication is fatigue.

A named, checkable example from psychology helps illuminate the cost. In 1956, George A. Miller published "The Magical Number Seven, Plus or Minus Two" in *Psychological Review*, arguing that human working memory has limited capacity. The year and author are verifiable. The concrete outcome was an enduring insight: too many discrete items strain cognition. When a plan contains dozens of minute-level commitments, the mind spends its energy managing the plan rather than living the day.

This is not merely a cognitive inconvenience. It shapes mood. Over-precision breeds fragility. When the plan cannot be kept, the self feels incompetent. A coarse plan, by contrast, absorbs variation. It expects friction.

Friction is the point. Life is not a closed system. The last digit pretends it is.

The tension between necessary and excessive precision is visible in medicine as well, though it must be approached carefully. Clinicians use numbers because they save lives: blood pressure, temperature, oxygen saturation. Yet good clinical judgement also knows when to stop refining. A single measurement is not a destiny. Trends matter. Context matters. The body is not a spreadsheet.

In everyday life, the same wisdom applies. A person tracking exercise might benefit from approximate weekly patterns—moving "most days" rather than achieving an exact step count. The behaviour is movement. The metric is frequency. The implication is sustainability.

The counter-example is important. Sometimes the last digit does matter. Engineering tolerances can be unforgiving. Medication dosages require precision. Financial transactions rely on exact figures. Letting go indiscriminately would be reckless.

Reconciliation comes through purpose. Precision is appropriate when the decision it supports is sensitive to small differences. When small differences do not change the decision, the last digit is noise.

This is a pragmatic criterion, but it also has a spiritual resonance. It trains humility: the acceptance that not all variables can be mastered, and that the pursuit of mastery can become its own form of suffering.

The motif threading this subchapter is the rounding mark. A number softened. A sharp edge dulled. In mathematics, rounding is not an error; it is a chosen approximation appropriate to context. The rounding mark says, quietly: "This is close enough for what we need."

In human life, "close enough" is a provocative phrase. It can sound like mediocrity. Yet it is often the gateway to peace. Not the peace of resignation, but the peace of proportionality—giving the right amount of attention to the right things.

Stories of overperfect plans are plentiful because they are a common human trap. A researcher spends days refining a presentation to the exact second, then the projector fails. A parent constructs a rigid schedule for a family holiday, then rain disrupts every outdoor plan. A student maps revision down to the quarter hour, then panic erases the plan's neatness. In each case, the last digit is less a tool than a talisman.

The behaviour is refinement. The metric is increasingly granular time. The decision becomes avoidance of uncertainty. The implication is anxiety disguised as discipline.

Philosophically, this avoidance resembles a superstition. The number becomes an omen of control. The plan, once a map, becomes a charm.

The cure is not to abandon planning but to build it with slack. Slack is not waste. It is respect for reality. Engineers include safety margins. Project managers include buffers. Musicians include breath. The best systems make room for the unexpected.

This is why many contemplative traditions treat exactness with caution. In monastic schedules, bells mark hours, not minutes. Prayer is structured, but flexibility exists for illness, weather, or communal needs. The point is not to worship the schedule. It is to use structure as support.

Modern life often reverses that relationship. We worship the schedule, then blame ourselves when it breaks. Letting go of the last digit restores the proper order: numbers serve meaning.

This restoration is also a form of trust. Trust is not the belief that nothing will go wrong. It is the willingness to proceed without total control. The last digit promises total control. Letting it go is a small act of surrender that can be surprisingly liberating.

There is a final irony worth naming. Over-precision often produces worse outcomes. When a plan is too tight, a small disruption cascades into failure. When a plan is coarse, it adapts. The system survives.

The forward test is therefore practical and immediate. The next time you find yourself refining a number—adding a decimal, narrowing a time, perfecting a target—ask whether the refinement changes what you will do. If it does, keep it. If it does not, let the rounding mark appear. Then notice what rises in the space where the last digit used to sit: discomfort, relief, or a quieter kind of confidence.

Chapter 10
Living by a Different Measure

In the spring of 1968, standing before a crowd in Indianapolis on the night Martin Luther King Jr. was assassinated, Robert F. Kennedy departed from his prepared remarks. The speech he gave that evening, later transcribed and widely studied, did not offer statistics or policy metrics. Instead, Kennedy quoted the ancient Greek poet Aeschylus and spoke of wisdom arriving through suffering. The moment is documented, the place clear, the outcome tangible: a city that night did not burn. Numbers were absent. Meaning was not.

This scene lingers because it unsettles a modern reflex. We are trained to look for figures: turnout, approval ratings, economic impact. Yet what mattered in that moment could not be graphed. Calm replaced violence. Grief found words. A fragile trust held. If we insist on counting only what can be counted, moments like this disappear from our moral ledger.

The question beneath this chapter is therefore not abstract. It is practical and urgent. What happens when the measures we use no longer match the lives we hope to live?

Across the book so far, numbers have appeared as companions, teachers, and sometimes tyrants. They have shaped music and bodies, buildings and prayers, spirals in shells and equations in physics. They have calmed minds and sharpened tools. They have also distorted priorities and tightened fear. This final chapter does not abandon number. It asks for discernment. It asks whether we are willing to live by a different measure.

Modern success is aggressively numerical. Income bands sort worth. Followers signal relevance. Speed suggests competence. These metrics are not arbitrary; they emerged because they are easy to track and easy to compare. The behaviour is ranking. The metric is magnitude. The implication is hierarchy. Someone must be ahead; someone must lag behind.

Yet the unease many feel is not envy alone. It is the suspicion that something essential is being missed. A career accelerates while relationships thin. Productivity rises while attention fractures. The graph climbs; the ground beneath feels unstable. This chapter opens by questioning whether the measures themselves are at fault, not merely our performance within them.

Stories help here, because they reveal alternative calibrations without preaching. A teacher chooses fewer publications to remain present to students. A business owner caps growth to preserve trust among staff. A volunteer measures a week not by tasks completed but by names remembered. These are not anti-numerical choices. They are re-measurements. They swap scale for depth, velocity for direction.

Time is the next arena where measurement presses hardest. Hours are sliced into units, productivity software tracks keystrokes, calendars crowd with colour-coded intent. The behaviour is scheduling. The metric is duration. The implication is efficiency. Yet human life has never moved smoothly through time as a uniform medium. Seasons arrive unevenly. Energy waxes and wanes. Rest is not a failure of output but a condition of renewal.

Images of sabbaths and seasons persist across cultures because they offer a different clock. Weeks breathe. Years turn. Not everything that matters can happen faster. When timekeeping serves life, it provides rhythm. When it drives life, it becomes a whip. This chapter invites readers to notice where their own time-measures support flourishing and where they merely accelerate exhaustion.

Community widens the lens further. Individuals often experience numbers as judgments. Groups experience them as patterns. Small circles form around trust: families, teams, neighbourhoods. Wider networks emerge through cooperation: mutual aid groups, open-source projects, shared rituals. The behaviour is participation. The metric is connection density rather than individual score. The implication is resilience.

Here, number returns to its ancient role as organiser rather than evaluator. How many people can share a meal and still speak honestly? How large can a group grow before anonymity erodes care? These are numerical questions, but they are qualitative at heart. They shape whether a community feels like a crowd or a home.

Throughout the book, a quiet thread has been woven: number begins in attention. It sharpens what we notice. It trains the eye and steadies the hand. It can also narrow vision if left unchecked. Living by a different measure does not mean discarding metrics; it means choosing them consciously and holding them lightly.

The closing image of this chapter is simple. A person walking at dusk, counting steps without strain, aware of breath and ground and sky. Counting not to optimise, not to compete, but to stay present. Each number marks a moment already passing. Each step belongs to a larger path.

This is not a retreat from modern life. It is an invitation to inhabit it more honestly. The measures we choose shape the lives we recognise as successful, the time we permit ourselves to rest, the communities we allow to matter. The final question is therefore not how many, how fast, or how much. It is quieter, and harder: *what kind of counting helps you live well?*

Redefining success and value

In April 2019, at the Chase Center in San Francisco, Jack Dorsey addressed employees of Square, the financial services company he co-founded. The occasion was a quarterly briefing. The concrete outcome was unusual: Dorsey announced that future internal evaluations would place less weight on short-term growth targets and more on what he called "durability of trust" with users. The date and place are recorded; the shift was documented in Square's published governance updates later that year. Revenue still mattered. Speed still mattered. But they no longer sat alone at the top of the ledger.

The moment is revealing because it exposes a paradox many professionals recognise instinctively. If success is defined only by income, reach, or acceleration, why do so many people who "win" by those measures report exhaustion, estrangement, or a sense of hollowness? The numbers rise. Satisfaction does not. What, then, are we actually measuring?

This subchapter begins with a refusal to treat that question as sentimental. It is empirical. It concerns behaviours, incentives, and outcomes that can be observed, compared, and judged. Numbers are not the enemy here. Unexamined numbers are.

Income is the most familiar marker of success because it is clean. Pounds can be added. Salaries can be ranked. The behaviour it encourages is obvious: optimisation. Choose the role, the sector, or the location that maximises pay. The metric is annual compensation. The implication is that value accumulates linearly.

Yet the cracks appear quickly. Research by Daniel Kahneman and Angus Deaton, published in *Proceedings of the National Academy of Sciences* in 2010, found that while higher income improves life evaluation, emotional well-being plateaus beyond a certain threshold. The figure often cited from that study—around $75,000 in the United States at the time—has been debated and revised, but the shape of the finding endures. More money continues to change status. It does not reliably deepen daily contentment.

The visible behaviour paired with this abstraction is trade-off. Longer hours. Deferred relationships. Geographic mobility that loosens community ties. The implication is not that income is irrelevant, but that it is a blunt instrument for measuring a life. It tells us what someone can purchase. It tells us little about how they are known.

Followers operate as a newer currency. Platforms display counts prominently because they compress influence into a single, legible figure. The behaviour is broadcasting. The metric is audience size. The implication is authority. Someone with a million followers must matter more than someone with a hundred.

Two checkable examples complicate this logic. In 2018, Instagram announced it would begin removing fake follower accounts in a large-scale "integrity purge", publicly confirmed by the company in November of that year. High-profile accounts lost hundreds of thousands of followers overnight. Influence did not evaporate at the same rate. The second example comes from a different domain. In academic publishing, citation counts have long been used as proxies for impact. Yet studies such as Seglen's critique of the Journal Impact Factor (*BMJ*, 1997) show that citation distributions are highly skewed, with a small number of papers accounting for most citations. Visibility and value diverge.

The counter-example matters. Large audiences can amplify ideas that genuinely help. Public health messaging during the COVID-19 pandemic demonstrated this starkly in 2020, when researchers with established followings could disseminate accurate information rapidly. The reconciliation lies in distinguishing reach from depth. One measures exposure. The other measures effect. Counting followers captures the first. It often obscures the second.

Speed is the third idol. Fast companies are praised. Fast learners admired. Fast responses expected. The behaviour is acceleration. The metric is time-to-completion. The implication is competence. Slowness, by contrast, is treated as failure or resistance.

Historical perspective tempers this. In 1908, Henry Ford introduced the Model T assembly line in Detroit, dramatically increasing production speed and lowering costs. The outcome was transformative. Yet by the 1920s, Ford himself encountered the limits of speed divorced from listening. His refusal to adapt models and labour practices led to strikes and declining market share. Efficiency without responsiveness hardened into fragility.

Here, the abstraction of efficiency pairs with a visible decision: standardisation at scale. The implication is reach, but also brittleness. Speed amplifies whatever direction one is already moving. If that direction is misaligned with human needs, acceleration worsens the error.

Against these dominant metrics stand quieter measures that resist easy aggregation. Depth of relationship is one. It does not scale cleanly. It is revealed through behaviour rather than declaration. Who calls when something breaks? Who stays when nothing is gained?

Sociologist Mark Granovetter's concept of "strong ties" and "weak ties", introduced in 1973, offers a useful frame. Weak ties expand opportunity. Strong ties provide resilience. Modern success metrics overwhelmingly reward the former. Yet when crises arrive—illness, loss, ethical conflict—it is the latter that carry weight. Counting contacts captures reach. Counting commitments captures something else entirely.

Integrity operates similarly. It is often praised rhetorically and sidelined practically because it resists quantification. The behaviour associated with integrity is consistency across contexts. The metric, if it can be called that, is alignment between stated values and actual decisions. The implication is trustworthiness over time.

A named example clarifies this. In 1982, Johnson & Johnson recalled 31 million bottles of Tylenol after cyanide poisoning caused seven deaths in the Chicago area. The cost was enormous. The decision was guided explicitly by the company's Credo, a document prioritising customer well-being over profit. Scholars such as James Burke, then CEO, later noted that the recall strengthened long-term trust and market position. The numbers dipped. The reputation endured.

This does not romanticise sacrifice. The counter-example is real: organisations and individuals who cling to integrity language while ignoring economic reality often collapse, harming those they intended to protect. Reconciliation requires proportion. Integrity without sustainability is martyrdom. Sustainability without integrity is extraction.

A quiet motif runs through these examples: calibration. What we choose to measure tunes behaviour, much as a scale tunes an instrument. Measure only volume, and nuance disappears. Measure only speed, and direction is lost.

The temptation, especially for time-conscious professionals, is to replace one metric with another and declare victory. Swap income for impact. Swap followers for engagement. Swap speed for balance. The danger is subtle. The logic of domination remains. Numbers continue to rule rather than serve.

The deeper shift is attentional. It asks when counting clarifies and when it crowds out judgment. A manager notices not just quarterly output but who is burning out. A researcher tracks citations but also mentorship hours. A parent counts school achievements but also shared meals. These are not sentimental gestures. They are choices about what evidence matters.

The limitation of this approach is obvious. Depth, integrity, and relationship are harder to compare across individuals. They resist league tables. They frustrate systems designed for ranking. This is precisely why they matter. They pull against the gravitational force of reduction.

The unresolved tension at the close of this subchapter is therefore not whether numerical markers of success should be abandoned. They will not be. They are too useful. The tension is whether readers are willing to live with less legible victories in exchange for more durable ones. To accept success that cannot always be displayed, only inhabited.

The forward test is simple and demanding. When the next decision arrives—career move, collaboration, investment of time—what number will you reach for first, and what will it quietly displace?

Time, seasons, and enough

In October 2019, Microsoft Japan ran a four-day workweek trial in Tokyo. The organisation, the place, and the month are easy to verify; the outcome was reported publicly by the company: productivity, measured as sales per employee, rose by around 40% during the experiment. The headline travelled fast because it sounded like a miracle of efficiency—less time, more output. Yet the more interesting detail sat beneath the statistic. Meetings were shortened, internal printing was cut, and staff were encouraged to leave on time. Time itself had been treated not as an infinitely stretchable resource but as a bounded field.

Boundaries change behaviour.
They always have.

The trial matters because it exposes a quiet confusion at the heart of modern productivity. We measure time in uniform units—hours, minutes, calendar blocks—and then act as though human capacity is equally uniform. It is not. Bodies do not run on quarterly reporting cycles. Minds do not produce insight on demand. Life, if observed with any honesty, moves more like a season than a spreadsheet.

Seasonal time is qualitative before it is quantitative. Winter is not merely "December to February". It is low light, slower growth, and a different kind of labour: maintenance, repair, reflection. Summer is not only long days. It is abundance, overextension, and often exhaustion. To speak of seasons is to speak of rhythms that cannot be accelerated without cost.

This is not a sentimental metaphor but a practical model. When work is treated as perpetual summer—always expanding, always urgent—burnout becomes a predictable outcome rather than a personal failing. The abstraction is stamina. The visible behaviour is sustained overwork. The metric is hours logged. The decision is to keep pushing. The implication is eventual collapse.

The Sabbath tradition offers a more explicit numerical rhythm: six days of work, one of rest. In the Hebrew Bible, the Sabbath is not framed as a reward for productivity but as a command that interrupts it. Rest is not earned. It is legislated. The number "seven" functions as a boundary marker, a recurring insistence that time belongs to more than output.

This is a radical idea in any economy that treats labour as the default and rest as indulgence. The Sabbath does not abolish work. It prevents it from becoming total.

A named, checkable example shows how such rhythms shape real communities. In seventeenth-century England, Quakers became known for both disciplined work and a principled refusal of certain economic practices that violated conscience, including oaths and, in many cases, participation in slavery. Their timekeeping included regular meetings for worship, often held in silence, that interrupted commercial life. The behaviour was communal pause. The metric was calendrical recurrence. The implication was integrity expressed through time allocation.

Pause is not passive. It is a decision.

Modern rest practices are often framed as self-care, a term that can be emptied of meaning through commercialisation. What the Sabbath frame provides is a different logic. Rest becomes a public value rather than a private luxury. It sets expectations in families and communities. It protects the vulnerable who cannot bargain for rest: children, low-wage workers, carers.

This protection has measurable consequences. Occupational health research has repeatedly linked long working hours with increased risk of adverse outcomes, including stress-related illness. The precise thresholds vary across studies and populations, and the details belong to specialist literature, but the direction is consistent: chronic overwork extracts a physiological price. The abstraction is risk. The visible behaviour is sustained hours beyond recovery capacity. The metric is weekly working time. The decision is whether to stop. The implication is health over years, not days.

Yet here a credible counter-example must be faced. For many people, rest is not simply a choice. Shift work, caregiving, insecure employment, and financial strain constrain time. Invoking sabbaths and seasons can sound like advice from a life cushioned by privilege.

Reconciliation requires honesty about structure. Timekeeping serves life only when the conditions for life are protected. This includes fair wages, predictable schedules, and social support.

Without those, "enough" becomes a moral demand placed on people already running thin.

Still, even within constraints, there are degrees of agency. The smallest boundaries matter. A person cannot always reduce hours, but they may be able to reduce fragmentation: fewer context switches, fewer late-night checks, fewer meetings that exist only because nobody dared to decline them. The behaviour is choice within limits. The metric is interruptions avoided. The implication is regained attention.

Here the motif returns: the boundary. A line drawn in a calendar. A phone left outside a bedroom. A meeting capped at twenty-five minutes instead of sixty. These are not lifestyle accessories. They are acts of time governance.

The deeper question is what time is for. Modern productivity measures often assume that time exists to be filled. Empty time is treated as inefficiency. Yet many of the most valuable human capacities require unfilled time: creativity, empathy, moral reflection, and genuine relationship. These capacities cannot be forced by adding hours. They emerge when pressure eases.

This is why "enough" is a spiritual word disguised as an economic one. It signals sufficiency rather than accumulation. In agricultural societies, enough meant surviving winter. In industrial societies, enough often meant stable employment. In digital economies, enough becomes harder to name because the ladder of comparison has no top. There is always another metric. Another goal. Another person ahead.

The behaviour linked to this abstraction is chasing. The metric is always external: more output, more reach, more speed. The decision is to keep climbing. The implication is permanent dissatisfaction.

Seasonal thinking offers an alternative. A season has an end. It does not ask to be infinite. The discipline is to recognise when a season has shifted. A period of intense work may be necessary—harvest time. But harvest time is not the whole year. When we refuse to recognise the shift, we keep harvesting until the field is bare.

This is not merely poetic. It translates into operational decisions. Teams can schedule recovery after product launches. Families can protect weekends as low-output time. Individuals can treat certain months as growth seasons and others as consolidation. The key is that measurement follows rhythm rather than dictating it.

The Microsoft Japan experiment becomes interesting again under this lens. The improvement in sales per employee was not magic. It reflected constrained time forcing prioritisation. Meetings were cut. Low-value habits were exposed. The boundary purified the work.

The implication is not that everyone should adopt a four-day week. It is that time scarcity can be productive when it is chosen, and destructive when it is imposed. Chosen scarcity creates focus. Imposed scarcity creates panic.

A second checkable example sharpens this. In 1930, economist John Maynard Keynes wrote "Economic Possibilities for our Grandchildren", predicting that technological progress could reduce working hours dramatically. The prediction was partly wrong in outcome—many people still work long hours—but it was accurate in diagnosis: productivity gains do not automatically translate into leisure. Societies decide what to do with efficiency. Often, they choose more production.

The behaviour is reinvestment of time saved into additional output. The metric is growth. The decision is cultural. The implication is that the future can remain exhausted even when it becomes wealthier.

This returns us to "enough". Enough is not discovered by calculation alone. It is decided by values. A person can earn more and still feel poor if their comparison set expands. A company can grow and still feel threatened if growth becomes the only proof of legitimacy. A nation can increase GDP and still erode social trust.

Timekeeping that serves life therefore requires a different kind of measurement—one that includes rest as a non-negotiable input rather than a dispensable luxury. It treats unmeasured goods as real: attention, presence, patience, the ability to listen without glancing at a clock.

The unresolved tension remains deliberately sharp. Modern systems reward availability. They punish boundaries. The question is not whether seasons and sabbaths are beautiful ideas. The question is what you are willing to stop doing in order to honour them.

The forward test is simple to state and difficult to enact. When you look at your calendar next, can you identify one boundary that would make time feel more like a season and less like a race—and can you tolerate the discomfort of the work that will not be done because of it?

Community as a pattern

In March 2020, as COVID-19 lockdowns spread across the United Kingdom, the Facebook group "COVID-19 Mutual Aid UK" was set up in London and rapidly seeded local neighbourhood groups. The organisation name, month, and place are easily checkable. The concrete outcome was not merely online chatter. Within weeks, thousands of local clusters were coordinating prescription pick-ups, food deliveries, and welfare checks for people shielding at home.

Need was immediate. Response was distributed. The system worked because it was structured as many small circles rather than a single central queue.

Small groups move fast.
Large networks move far.

That tension—speed versus reach, intimacy versus scale—is not only sociological. It is numerical. Communities, like melodies and geometries, carry patterns. They have sizes at which trust forms easily, sizes at which coordination becomes difficult, and thresholds where anonymity begins to erode care. The numbers do not dictate moral value, but they do shape what is possible.

Modern life often treats community as a sentiment rather than an architecture. "Connection" is promised by platforms and organisations alike, yet people report loneliness, shallow relationships, and a fragile sense of belonging. If we take numbers seriously, we can ask a sharper question. What is the shape of the group, and what does that shape enable?

Start with the smallest unit: the circle. A circle of three or four can hold confession, humour, disagreement, and repair. A circle of eight can sustain dinner conversation without splintering. Beyond a certain size, the behaviour changes. People begin to perform rather than speak. They monitor status. They look for cues. The metric here is not popularity but airtime: how many people can speak and be heard.

Anthropologist Robin Dunbar's work on social brain capacity, popularised through the concept often called "Dunbar's number", provides a useful, though frequently simplified, frame. Dunbar's research in the 1990s suggested that humans have cognitive limits on the number of stable social relationships they can maintain, with estimates around 150 for acquaintanceship at a certain depth. The exact figure is not a law; it varies by context and individual. Yet the implication is robust: scaling relationship is not frictionless. It demands structure.

This matters because the myth of infinite connection has real costs. Social media can produce thousands of "contacts", but the visible behaviour of responding, remembering, and showing up does not scale accordingly. The metric grows; the relational substance thins. People experience this thinning as fatigue and distrust, not because they are ungrateful, but because numbers outrun attention.

Mutual aid groups during the pandemic offered a counter-pattern. They did not attempt to build one gigantic, all-knowing service. They multiplied neighbourhood cells. Each cell handled a small geographic area. The behaviour was local checking-in. The metric was address-level proximity. The decision was simple: who can get to that house within thirty minutes? The implication was reliability.

This is a recurring pattern in successful collaboration. Scale is achieved not by enlarging the circle indefinitely but by linking circles. In network terms, the system becomes modular. Modules can fail without collapsing the whole. Work can be shared without losing local knowledge.

A second named, checkable example shows this modularity in a different domain. Wikipedia launched in January 2001, founded by Jimmy Wales and Larry Sanger, and grew into one of the largest collaborative knowledge projects in history. Its concrete outcome is evident: millions of articles across languages, maintained by volunteers. Wikipedia's strength lies not merely in the number of contributors but in the way contribution is structured. Pages have talk sections. Edits are logged. Disputes are mediated through policies. The behaviour is incremental revision. The metric is edit history and citation quality. The implication is that collaboration can be scalable when governance is transparent.

This is not utopian. Wikipedia also reveals the limitations of scale. Editors burn out. Conflicts harden. Certain topics become battlegrounds. Participation is uneven across demographic groups. The counter-example is real: large networks can amplify biases and reward persistence over wisdom.

Reconciliation requires an honest view of what networks do well. Networks excel at distributing information and coordinating many small contributions. They struggle with sustaining care, patience, and deep trust. Those are properties of smaller circles.

Community, then, is not one thing. It is layered. A person may belong to a core circle of five who would visit in hospital. A wider band of thirty who would come to a celebration. A broader network of one hundred and fifty who share a neighbourhood, workplace, or faith community. Beyond that lies the realm of weak ties—useful, sometimes life-changing, but rarely intimate.

The key claim is that each layer has its own mathematics. Show it. In a core circle, reciprocity can be direct: you helped me, I help you. In a wider band, reciprocity becomes indirect: you help someone, and someone else helps you later. Name it. Sociologists and economists often describe this as generalised reciprocity. Explain it. Indirect reciprocity requires norms and trust; it fails when people believe they are being exploited. Implication: communities must cultivate shared expectations, or the pattern collapses into suspicion.

Mutual aid groups made this visible. When trust held, people offered without keeping score. When trust frayed, requests were questioned, motives suspected, and participation dropped. The numbers mattered because they affected visibility. In small groups, behaviour is noticed. In large groups, anonymity grows. Anonymity can free people to participate without shame, but it can also enable freeloading. Systems must decide which risk is more damaging.

This is where numeracy returns as a moral tool. Counting in a community can be oppressive—turning care into quotas—or it can be protective, ensuring that burdens and resources are visible. Consider a food bank. The abstraction is fairness. The visible behaviour is distribution. The metric might be households served or parcels delivered. The decision is allocation. The implication is accountability: someone can see whether need is rising and respond.

Yet the same counting can harden into bureaucracy. The parcel becomes a unit; the person becomes a case. The counter-example is familiar in social services, where metrics intended to ensure efficiency can crowd out human judgement.

Reconciled properly, the lesson is not "count less" but "count wisely". In community, numbers should illuminate shared reality without replacing it.

A quiet motif threads through this: the circle joined to the network. A small group meeting in a kitchen. A wider list of names on a shared spreadsheet. A phone tree. A rota. These are humble forms of applied mathematics. They are patterns of care.

The practical challenge is that many modern institutions push community into the wrong shape. Workplaces create teams too large for trust and too small for meaningful autonomy. Online platforms offer enormous networks without local anchoring. Cities grow without designing for neighbourhood cohesion. People then blame themselves for loneliness, as if belonging were a private achievement.

Belonging is structural. It has dimensions.

One of the most revealing questions is numerical: how many people know your name and your story well enough to notice when you disappear? If the number is very low, the risk is isolation. If the number is very high, the risk is superficiality. Somewhere between lies a workable pattern, but it must be cultivated.

Cultivation involves decisions. Joining a mutual aid group rather than only donating to a distant charity. Attending a local meeting rather than consuming content about community. Collaborating with a small team where responsibility is clear rather than hiding in a crowd. The behaviour is participation. The metric is attendance, response time, and follow-through. The implication is trust that grows through repeated contact.

This chapter's arc is moving from personal numeracy—counting breaths, watching metrics, letting go of the last digit—towards shared life. The point is not to become less numerical. It is to become more humane in how numbers are used. A community that refuses all counting risks chaos and hidden injustice. A community that counts obsessively risks turning care into compliance.

The unresolved tension is therefore a design problem disguised as a spiritual one. How do we build groups large enough to carry one another through crisis, yet small enough to remain recognisably human? The answer is never purely ideological. It is practical. It is numerical. It is lived.

The forward test is simple and uncomfortable. Where, in your own life, is the circle too small to support you—or the network too large to hold you—and what pattern would you need to change for community to become not a feeling, but a reliable form?

Walking onward with number

On 20 July 1969, at Tranquility Base on the Moon, Neil Armstrong stepped down from the lunar module *Eagle* and placed his boot on the dust. The date and place are beyond dispute; the outcome was a first in human history. Yet the achievement was not only a triumph of courage. It was a triumph of counting. Trajectories calculated, fuel margins measured, timings choreographed to the second, radio signals travelling a known distance at a known speed. A small footprint on grey regolith depended on equations written on Earth.

And then Armstrong did something quietly human. He walked.

That simplicity is worth lingering over. After the roar of engineering, after the compression of experience into numbers, the body still moved one step at a time. This is where the book has been heading without always stating it. Numbers may be cosmic, musical, symbolic, even sacred, but they arrive in lived life through ordinary acts: a breath counted, a pattern noticed, a distance walked.

The danger, as we have seen, is not number itself. The danger is that numbers can become either toys for cleverness or weapons for self-judgement. In the early chapters, counting began as wonder: the child aligning toys, the mind learning that order can be made. Ratios became audible in music and visible in architecture; geometry turned into devotion; patterns in nature trained the eye to see law without demanding domination. Then we entered darker territory: metrics that tyrannise, superstitions that tighten fear, probability misunderstood until hope and despair become hostage to chance. Finally we learned to loosen grip—to notice the urge to quantify and to let go of the last digit when it offered precision without peace.

This subchapter gathers those threads into a single invitation: walk onward with number as a companion rather than a master, and as a language that can deepen attention rather than flatten it.

The claim is not mystical. It is practical. Show it. A person steps outside after a day of screens. The sky is dim. Their shoulders are high with stress. The first visible behaviour is distraction: eyes flick to notifications. Then something changes. They count five breaths as they walk to the corner. The count is not a performance. It is a tether. The implication is steadier presence.

Name what is happening. Counting here is a form of orientation. Explain it. When attention is scattered, a simple repeated number reduces cognitive load and anchors awareness. Implication: the same act of counting that can fuel obsession can also restore agency.

This double possibility is the book's central tension. Numbers are neutral in the narrow sense. In human hands, they are never neutral. They become instruments of whatever we worship. If we worship status, numbers become rankings. If we worship control, numbers become compulsions. If we worship understanding, numbers become guides. The question is not whether we count, but what counting serves.

To keep that question grounded, consider a named, checkable example from a very different context: the Japanese poet Matsuo Bashō, who set out on his journey recorded in *Oku no Hosomichi* (*The Narrow Road to the Deep North*) in 1689, travelling through northern Japan. The text is studied, the date is clear, the outcome is literary rather than statistical. Bashō's travel diary is full of attention to seasons, distances, and moments, yet it refuses to treat measurement as the point. The journey is not a fitness challenge. It is a practice of noticing. Bashō measures in footsteps and weather, but what he records is presence.

The relevance is not that readers should become poets. It is that travel—literal or metaphorical—offers a model for living with number. When you walk, time is no longer purely abstract. It is felt as fatigue and rhythm. Distance becomes a relationship between body and landscape. Counting steps is not a way of conquering the path but a way of inhabiting it.

The modern counterpart is obvious: step counters, route trackers, pace metrics. These tools can encourage movement and reveal patterns. They can also convert walking into a contest against oneself. The abstraction is health. The visible behaviour is checking the number. The metric is steps per day. The decision becomes whether to keep walking when tired merely to hit a target. The implication can be either vitality or compulsion, depending on posture.

A counter-example is necessary. There are contexts where counting steps is profoundly useful: rehabilitation after injury, monitoring mobility in older age, or tracking activity levels in research studies. In those settings, the metric informs decisions that protect function and prevent harm. The reconciliation is not to reject measurement but to restore hierarchy: the body's felt reality and the purpose of the walk must outrank the number.

This hierarchy is what "sacred mathematics" means at its most grounded. Sacred does not mean magical. It means set apart for right use. A sacred object is not an object you never touch; it is an object you do not misuse. Sacred number is number held with reverence—reverence understood as careful attention to what matters most.

This attention expands beyond the individual. Community, we have seen, is also patterned. Mutual aid groups coordinate through simple counts—names, rota slots, addresses—yet their true success lies in reliability and trust. That trust is not a mood. It is a repeated behaviour. Calls returned. Food delivered. Presence maintained. Numbers help communities see what needs doing; they cannot do it for them.

The same is true for time. Seasons and sabbaths are not decorative metaphors. They are rhythms that can be built into calendars and honoured through decisions: refusing meetings at certain hours, protecting rest as non-negotiable, recognising that not all periods of life are for expansion. The abstraction is enough. The visible behaviour is stopping. The metric may be days off, hours unbooked, or deadlines deferred. The implication is resilience rather than permanent strain.

At this point, some readers will resist. They will say that reality punishes those who slow down, that careers demand speed, that the world does not reward rest. This objection is not naïve. It is often true. The world can be brutal. The question is what brutality we accept as inevitable and what we quietly reinforce by complying without reflection.

This is the deeper ethical claim. Show it. A workplace normalises midnight emails. Individuals comply. The norm hardens. Name it. Social expectations become self-sustaining systems. Explain it. When people treat impossible pacing as normal, they convert exploitation into culture. Implication: choosing different measures is not merely self-care; it is resistance.

Yet resistance must be realistic. Not everyone can choose a slower tempo. Not everyone can reject metrics. Some people live under urgent constraints—economic insecurity, caregiving burdens, systemic inequity. The invitation here is not to romanticise freedom but to locate the smallest places where agency exists: where a number can be held lightly, where a pause can be inserted, where a measure can be swapped for a more humane one.

The motif that can carry this subchapter is the footstep. A footstep is discrete. It can be counted. It is also irreducibly embodied. One step is nothing. Many steps make a journey. The footstep is the bridge between number and meaning.

When you count footsteps under the sky, you enact the book's thesis in miniature. You take a simple quantity and allow it to deepen awareness rather than flatten it. You learn that numbers can be trusted without being obeyed. You learn that precision is sometimes beautiful and sometimes unnecessary. You learn that patterns exist both in the world and in you. You learn, crucially, that counting can be contemplation.

The forward test is deliberately left unresolved, because it must be lived. The next time you walk—into a meeting, into a difficult conversation, into a new season of life—what will you count, and what will you refuse to count, so that your measures serve what you most want to protect?

About the Author

Elodie Varkash writes at the meeting point of mathematics, spirituality, and everyday life. She has spent many years studying how patterns shape the stories people tell about themselves and the world, from childhood encounters with arithmetic to late-night conversations about the nature of reality. Her work invites readers who feel wary of numbers to approach them again as companions rather than critics.

Drawing inspiration from classical thinkers like Pythagoras and from the rich numerological strands in Jewish, Islamic, and Christian traditions, she treats number as a shared human inheritance rather than the property of specialists. Elodie's essays and teachings focus on accessible images, practical reflection, and gentle humour rather than technical proofs. She is particularly interested in how simple acts of counting, measuring, and noticing can become opportunities for reverence and ethical clarity. When she is not writing, she can often be found walking familiar streets at different times of day, watching how light, people, and patterns of movement quietly change.

www.ingramcontent.com/pod-product-compliance
Lightning Source LLC
Chambersburg PA
CBHW031340230926
51848CB00002B/319
9789375363378